AF714549

Transforming Innovations in Africa

African Dynamics

VOLUME 11

The titles published in this series are listed at brill.com/ad

Transforming Innovations in Africa

Explorative Studies on Appropriation in African Societies

Edited by

Jan-Bart Gewald
André Leliveld
Iva Peša

BRILL

LEIDEN • BOSTON
2012

Cover illustrations: Top right: "A single wheeled "garetta" southern Zambia circa 1910", photo courtesy of the National Archives of Zambia.
Bottom right: "Toyota HiAce built circa 1986 and still going strong as a minibus in 2010", Mali 2010, photo by Jan-Bart Gewald.
Bottom left: "Women bringing water to a steam traction engine southern Congo 1915", photo courtesy of the National Archives of Zambia.
Top left: "Craftsman regrooving millstones with an improvised lathe", Mali 2010, photo by Jan-Bart Gewald.

Library of Congress Control Number: 2012951191

This publication has been typeset in the multilingual "Brill" typeface. With over 5,100 characters covering Latin, IPA, Greek, and Cyrillic, this typeface is especially suitable for use in the humanities. For more information, please see www.brill.com/brill-typeface.

ISSN 1568-1777
ISBN 978-90-04-24523-5 (hardback)
ISBN 978-90-04-24544-0 (e-book)

This book is printed on acid-free paper.

Contents

Maps, figures, tables, boxes and photos

Maps

Figures

Tables

Boxes

Photos

Acknowledgements

The chapters in this book are based on papers presented at the workshop entitled 'Transforming Innovation in Africa' that was held at the African Studies Centre in Leiden from 26-28 October 2011. Many people have contributed to the production of this book. First and foremost, we would like to thank the authors who reacted so positively to our invitation to submit and present a paper at the workshop and who later wrote these up as chapters for this volume. We appreciated their enthusiasm and constructive criticism.

We are grateful to the staff in the secretariat at the African Studies Centre for their help with the organization of the workshop and for their logistical support. We would also like to thank Ann Reeves for her careful editing of this volume, Mieke Zwart and Dick Foeken who did the layout and Nel de Vink for producing the maps.

1

Introduction: Transforming innovations in Africa; explorative studies on appropriation in African societies

Jan-Bart Gewald, André Leliveld & Iva Peša

> *make changes in;* ***innovation*** *(noun);* ***innovative*** *(adjective);* ***innovator*** *(noun);* ***innovatory*** *(adjective). {Latin (novus new)}* *(Allen 1984: 380)*

The twentieth century saw an 'upsurge of enthusiasm...for inventors and innovation, for the mass consumer products of industrial technology' and, ideologically, innovations were overwhelmingly coupled with visions of 'progress' and 'improvement' (Adas 1989: 402). Within the context of colonialism and post-colonialism, it was all too often assumed that innovations developed in Europe or the West could be transplanted wholesale and unchanged into Africa. African societies and cultures have always interacted with the outside world and there has been a frequent exchange of innovations between Africa and the rest of the world, with external innovations (i.e. those developed elsewhere) being introduced, appropriated and transformed in Africa. This volume contends that innovations have never simply been introduced and imposed from outside the continent but that they have always been actively transformed in their appropriation by the receiving societies and cultures. The chapters in this book highlight the agency and creativity with which external innovations have been transformed in the process of their appropriation by African societies and cultures.

Innovation is a term that evokes a multitude of ideas and emotions, and is positively associated with 'modernity', 'development' and 'progress'.

Economically, intellectually and socially, innovation is considered to be an important element in societies, if only because it has transformed them, for better or worse. In economics, innovation is seen as a central driver of economic growth, development and job creation, and the source of improvements in the quality of people's everyday lives. It is associated with events such as the discovery of fire, printing techniques, double-entry bookkeeping, the invention of electricity, the steam engine, and digital technology. In addition, it is assumed that innovation necessitates a particular institutional setting (states, universities, business enterprises, etc.) without which the development of sophisticated scientific knowledge and technology could not take place.

In the past three hundred years, the dominant discourse that has developed regarding Africa is one that sees the continent as being 'backward', both technologically and with reference to innovation. Africa is regarded as the continent where innovation 'failed', where the wheel, literacy and industrialization were all 'late'. In the early 1970s, anthropologist Jack Goody argued that the absence of the wheel in Sub-Saharan Africa throughout much of the pre-colonial period made it impossible to develop animal, wind and water power, and thus delayed agricultural development, transport, military advances and even state formation in Africa (Goody 1971: 26-27, 47-48, 75-76). Similarly, innovations that did catch on were often considered to be externally generated. Knowledge of ironworking and agriculture, for example, has long been assumed to have 'diffused' to Sub-Saharan Africa from regions that were perceived as being more 'developed', such as the Middle East or North Africa, whereas only colonial domination brought about advances in industrial technology.[1] The lack of innovation in Africa in the above sense is often considered one of the major constraints for growth and development in the region. That Africa is a continent lacking innovation or innovative capacity is an image that is strengthened by the annual Global Innovation Index that aims to capture the richness of innovation in societies and ranks 125 countries accordingly (Dutta 2011). The 2011 index included 27 African countries, with South Africa and Mauritius ranking highest (50th and 53rd respectively), followed by Tunisia (in 66th place) and Ghana (70th). However, African countries occupied 17 of the bottom 25 places (with Sudan and Nigeria in 124th and 125th positions respectively). Another recent report on innovation and productivity in Africa concluded that 'innovation is the

[1] Such views are critiqued in many major works on African history, such as Iliffe (2007).

main driver of economic growth but the capacity to innovate is quite low in most African countries, both in the private and in the public sector' (Wolf 2007: Abstract).

Exactly what innovation is, however, remains unclear. Its definition is often highly value laden, depending on the context in which the concept is being used. And as Hall (2010: 2) observes: 'it shares with many other key concepts the paradox that once its importance is recognized, its meaning seems to drain away'. The concept of innovation runs the risk of being overused to the point that it becomes the signifier of nearly everything new, and consequently of almost nothing (*Ibid.*). In this sense, the dictionary definition of 'innovate' and 'innovation' at the start of this introduction does not take us very far. Hall (*Ibid.*: 3) also notes that the celebration of innovation in much of the literature as the key component in creativity and entrepreneurship, and thus presenting it as something positive, is unjustified. Innovation is not inherently beneficial. Hall (*Ibid.*: 3-4) gives the example of fast-food franchises that have been highly innovative in developing cheap products that are high in carbohydrates and unsaturated fats and popular with small children. Their products have been very successful but have unintentionally placed a direct burden on public-health systems and have reduced life-expectancy rates as a result of increasing levels of obesity. The value-laden nature of the concept itself also influences the way we view innovation or the perceived lack of it in Africa.

This volume considers innovation as a process by which ideas are transformed into practice (Hall 2010: 2). And more often than not, innovation comes from the cumulative effect of implementing small-scale ideas over prolonged periods of time. It is a process that encompasses the acts of numerous individuals, not only the original inventors but also the producers, consumers and middlemen that transmit and operationalize the innovations, making them acceptable to society. Innovation here thus refers to processes of invention, adoption, adaptation, appropriation and transformation. How these processes develop and their outcomes depend on the context in which they take place. Innovations are shaped by societies and cultures, in particular by institutions that can be defined as durable systems of established and prevalent social rules that structure social interactions and mould individual habits. Hodgson (2006) also argues though that individual habits of thought are not passive entities but constructs that are continuously and often repetitively invoked in response to contextualized stimuli. In his view therefore, institutions mould an individual's habits but still leave the individual room to manoeuvre and deliberately choose a different action that may lead to institutional change as well. How processes of innovation develop and their results depend, among other things, on this circle of mutual interaction and interdependence between individual habits and institutions. Such interaction takes place over time and the historical description and analysis of innovation processes in Africa is therefore a central issue in this volume. It considers the way that ideas, values and knowledge that already exist in a given society shape the course of innovation processes.

In addition to innovations initiated and developed on the continent, Africa abounds with examples of innovations that were initially envisaged, developed and/or designed outside the continent yet came to be unexpectedly and innovatively transformed in Africa. These external innovations might be of a material (the motor vehicle, the mobile phone, bicycles, firearms) or an immaterial nature (judicial and state systems, insurance schemes, educational curricula). The chapters in this volume specifically seek to deal with the transformation in Africa of innovations of both a material and an immaterial nature, innovations that initially originated and were developed outside of Africa and were introduced at a later stage. They explore how external innovations, i.e. products, services, technologies, institutions and processes, that are the outcome of innovation processes outside Africa, embark on new, internal innovation processes in African societies and cultures in such a way that they become acceptable to local conditions, expectations and demands. As will be shown, the result of this incorporation of external innovations in internal innovation processes is different from that which had originally been envisaged by those who introduced the innovation in the first place (see in particular the chapters by Yusuf and van Walraven, this volume). In a number of cases, the innovation was transformed unintentionally by its users, or was even rejected outright (see, for example, the contributions by Owuor & Foeken and Dekker & Kinsey, this volume).

The striking example of the Toyota Hi-Ace minibus and its utilization in and across Africa illustrates this point. Travelling by public transport in Africa, it becomes apparent to many observers that their chosen form of transportation has undergone certain modifications that are peculiar and particular to local conditions. From the souped-up Zola Toyota Hi-Ace minibuses of suburban South Africa to the barely functioning *Matatus* in East Africa and the slogan-emblazoned *Trotros* of Ghana, these Toyota Hi-Aces are being used in ways and conditions that could never have been envisaged by Toyota's designers and automotive engineers in Yokohama. Instead Hi-Aces have been adapted and transformed to suit local conditions. Be it macho prowess and/or whitewall tyres in Johannesburg, the constraints of limited budgets and cut-throat competition in Nairobi or the desire to put across a message in Accra, the epitome of Japanese engineering prowess has been transformed in each case in unique and unanticipated ways.[2] What holds true for Toyota Hi-Aces in Africa equally holds true for all other material forms of technological innovation (see in particular Bellucci & Zaccaria, this volume), be it a bicycle (MacCracken 2011), an assault rifle (Macola 2010), agricultural crops (Richards 1985) or mobile technology (de Bruijn *et al.* 2009).

In a similar vein, what applies to the material also applies to the immaterial forms of innovation used in Africa. Examples of this would include the concept of the nation-state as it developed in Europe following the Treaty of Westphalia, and which was introduced into Africa and deployed in the late colonial period in the run-up to independence. The imposition and continued functioning of the nation-state in Africa has led to the development of state and legal systems that are peculiar to Africa, and the subject of substantial debate that has littered African Studies with terms that have an illusory meaning, be it 'fragile states', 'façade states', 'vampire states', 'failed states', 'staged states' or 'neo-patrimonial states', to name but a few.[3] Apart from the nation-state, there are other forms of immaterial innovation that have their origins outside Africa yet were adopted and developed on the continent itself (for borders, see van Walraven, this volume). The film form that developed in Nigeria and has become known as Nollywood is a spectacular example of this. So too are the health insurance schemes, banking products, educational curricula, forms of broadcasting, and ideologies and religions that all have their roots outside Africa but have gained a specific form and function in Africa.

This volume presents the results of a three-day workshop hosted by the African Studies Centre in Leiden from 26-28 October 2011. What is striking is the strongly historical approach in many of the papers submitted. This in itself is not surprising as the transformation of innovation and innovation processes in and of themselves imply a diachronic dimension. However more than half of the chapters, although they may be dealing with present-day issues, consciously have the late colonial period, i.e. fifty years ago, as the starting point of their analysis. In addition to

[2] For examples and case studies of these local transformations, see Gewald *et al.* (2009).
[3] See also Bayart (1989) and Bayart *et al.* (1999).

their broad historical interest, the chapters deal with the transformative aspects of innovation for the social, political, cultural and economic conditions of societies and cultures, as well as the inverse, namely how the social, political, cultural and economic conditions of African societies have transformed innovations in Africa. The dialectical relationship that exists between African societies and innovation, with each influencing and transforming the other, is an important aspect in many of the chapters.

The chapters can be broadly divided into those that deal with innovations of a material nature and those innovations that are of an immaterial nature. However crude Marxism has taught us that there is a dialectical relationship between the material and the immaterial, and that any transformations in the material base of a society, the 'means of production' (the introduction of new tools or new factories) in and of itself immediately leads to changes within the 'relations of production' within a society. The introduction of a material innovation thus leads to changes within the immaterial and *vice versa*. Building on this, Social Construction of Technology (SCOT) theory shows that transformations in the material also lead to transformations in the immaterial domain, and *vice versa*. It is a central tenet of SCOT theory that any given technology only acquires its function within a specific social and cultural context. Thus different social groups can construct radically different meanings for a technology (technology's interpretive flexibility). As everyone who has attended school will confirm, a pencil does not in and of itself carry its function as a writing implement: it can be used to drum on a table, poke holes in a piece of plastic or stab an unsuspecting victim. It does so much more than merely transcribe a teacher's ABC onto a piece of paper.[4] It is not feasible therefore to divide the chapters up in terms of the transformation of the material from the immaterial concerning results, for any innovation within the material realm leads to a transformation in the immaterial realm. It is, however, possible to divide the chapters up in terms of those in which the innovation was envisaged within the material realm (the motorcycle or pump for example) from those where the innovation was envisaged in the immaterial realm (forms of belief and of governance). This division has been applied throughout the volume, although the editors are well aware of the dialectical relationship that exists between an envisaged material innovation and the immaterial transformation that results.

[4] Pinch & Bijker (1984) saw users as a social group that played a part in the construction of technology. See also Oudshoorn & Pinch (2003).

Research conducted within Technology Studies confirms the findings of Africanists – anthropologists, historians, sociologists, political economists and human geographers alike – that modernization is a contested process. Different social groups within a society appropriate and transform innovations in different ways as there is no unilinear appropriation of innovation. Complex and mutual interaction between the innovation, social groups and intergroup power relations can be observed. Ronald Kline's words on the electrification of the American Midwest are as true for America as they are for Africa: 'farmers contested efforts to urbanise rural life by resisting each technology, then weaving it into existing cultural patterns to create new forms of rural modernity' (Kline 2003: 52). The work of Kline & Pinch (1996), which showed how the Model T Ford was appropriated and redesigned by American farmers, rings particularly true for anyone who has ever conducted fieldwork in Africa. They described how American farmers used the Tin Lizzie, as it was affectionately known, as a stationary power source to drive everything from washing machines to lathes (Kline 2000). Innovations can be transformed in multiple and unexpected ways.

Innovation processes do not necessarily entail the further development of things, in the sense of further additions or attributes. Innovations envisaged within the immaterial realm, such as forms of belief or of social organization, would also include differing forms of envisaged economy, for example the market economy versus the planned economy, or Anglo-Saxon versus Rhineland business models. Within economics, the concept of Frugal Innovation has been coined to describe the way users and consumers are related to producers. The well-known example of the African market women who buy washing powder in bulk and sell small repackaged units for a profit contributed to market-leader Unilever anticipating this behaviour and introducing small 10-gram sachets (see the contribution by van Beers, Knorringa & Leliveld, this volume). Enterprises that develop frugal innovations take the needs of poor consumers as their starting point and work backwards so that, instead of adding ever more attributes and additions, they strip products down to their bare essentials. Jeff Immelt, CEO of General Electric, and Vijay Govindarajan of the Tuck Business School have dubbed this 'reverse innovation' (Govindarajan & Trimble 2012). Others call it 'frugal' or 'constraint-based' innovation. It is not just about redesigning products but involves rethinking entire production processes and business models. Related to this is the realization that, if profits are to be maintained, multinational corporations have to take into account the needs and wishes of the 'Bottom of the Pyramid', a concept that has become mainstream in economics of late (Prahalad

2012; Prahalad & Hart 2002). The sale of mobile-phone credit that allows a single SMS message to be sent is unique to Africa and reflects this desire for profit that can be obtained from the Bottom of the Pyramid. Another example is M-Pesa in Kenya (see Mwangi & Rutten, this volume).

It is not the claim or purpose of this volume to exhaustively expand on what the theoretical nature of innovation in Africa actually means and entails.[5] Nevertheless, a number of the case studies explicitly highlight theoretical concerns that call for further research (particularly those by Dobler and van Dijk, this volume). While some of the authors discuss analytical issues, others have adopted a purely empirical approach. Our main objective was to present case studies from different disciplinary backgrounds and to showcase a bewildering and fascinating spectrum of innovation processes with sometimes surprising outcomes in African societies and cultures. The case studies demonstrate the depth and richness of innovation in African societies, even if this is not the same richness as that defined in the Global Innovation Index. The richness described in the contributions in this volume is the result of authors taking a specific perspective on what innovation entails, namely focusing on innovation as an incremental process. This view of innovation and the outcome and values that it ultimately embodies can only be understood by analyzing the context in which processes of innovation take place and by taking a historical perspective.

In this context, we contend that although an invention might be seen as a single act, it rarely is, if only because an invention involves prolonged prior preparations in knowledge and skills as well as subsequent perfections. Innovation is *always* a process and one that involves numerous stages from its invention, introduction, adaptation, transformation, acceptance (or rejection) and appropriation. These stages together form a distinct process of innovation. It is the wide variety of innovation processes that is explored in this volume. The fact that the authors all see innovation as a process means that they have taken a diachronic approach in their contributions. Innovations are brought about by a multitude of actors (producers and users) and can have a variety of different outcomes. We have therefore considered questions such as: What factors contribute to innovations and their outcomes? What kinds of power relations underlie innovations? Can existing power relations in a society be altered by

[5] This has been done most lucidly by Adas (1989) and Austen & Headrick (1983).

an innovation? In answering these questions, we highlight the fact that innovations do not have the same effects on all groups in a society. While an innovation may be beneficial and prove empowering to some, they can be detrimental to others. Some groups might resist an innovation, give it new meaning, create a parallel structure or use the innovation as a means of social mobility or even emancipation (see the chapters by Nugent, Peša and Gewald, this volume).

The following provides a brief introduction to each of the papers, which have been grouped according to themes: socio-economic, institutional, socio-cultural and socio-technical innovations. Innovations encompass social processes in their entirety and cannot be confined to any single sphere. Nevertheless, it is possible to identify the realm from which an innovation has originated in order to question whether its origins have influenced the subsequent innovation pathway and its eventual outcome. Tracing these distinct innovation pathways through case studies can provide interesting parallels between the cases and offer a tool for future model-building or theoretical endeavours.

Socio-economic innovations

The contribution by Paul Nugent traces the history of the wine industry in South Africa. Of late, South African wines have acquired a recognized standing and a large international following. During the twentieth century the country's wine industry underwent numerous changes and modifications in an effort to fulfil a wide variety of, at times, competing demands. This chapter describes and analyzes what it was that determined and influenced the production of wine in South Africa at this time. Nugent provides the reader with a detailed overview of the way the innovation (in this case, wine) was appropriated and transformed in South Africa in such a manner that distinct South African wines and a wine culture have emerged.

In his historical survey, Gregor Dobler concentrated on the establishment and development of shops and shop keeping in Ovamboland in northern Namibia between 1925 and 1980. Constrained by a series of specific laws that restricted the setting-up of trading stores in Ovamboland, Dobler traces the development of the first independent privately owned stores, outlining the increasing differentiation between single-shelf shops and successful supermarkets. However, the fact that the area was a homeland that was governed and subject to a myriad of South African racial rules ensured that 'shops lived on outside rents and in a perversely protected environment' (Dobler, this volume), which was ultimately not sustainable. With changes to apartheid legislation, these local shops could not compete with mainstream companies and vanished into oblivion.

Unilever's struggle to gain and retain its place in the 'Bottom of the Pyramid' through 'frugal engineering' is illustrated in the chapter by Cees van Beers, André Leliveld and Peter Knorringa. Building on standard economic practice, market traders in Africa buy soap powder in bulk and repackage it in small sachets for sale at a small profit. In parts of Africa, Unilever, as the producer of OMO washing powder, was the market leader and commanded a sizeable segment of the market. Indeed, in some places all blue washing powder has come to be known by the generic term 'OMO'. However, some of the washing powder sold this way did not maintain the quality standards demanded by Unilever who, in an attempt to retain access to the market and its brand name credibility, sought to 'reverse innovate' its sales of OMO by producing small sachets of washing powder for the African market. These were to compete with the manually repackaged washing powders that were being sold as OMO. This is a case where a multinational attempted to copy and improve on an innovation initiated by African consumers.

The chapter by Moses Mwangi and Marcel Rutten on M-Pesa in Kenya is equally an example of a socio-economic internal mobile technology innovation that is building on an external innovation. The study, undertaken among the Maasai pastoralists of Kajiado County, sheds light on the impact of a mobile-phone-based money system (M-Pesa), and shows that technology has proved to be a way of overcoming some of the obstacles that block the rural poor's access to financial services. The mobile-money system has changed the way people manage their money and also how they organize their livelihoods. Women have been empowered and can now manage their own income from various sources without any fear of interference from the men in their community.

Institutional innovations

In his contribution on the history of the mass political movement Sawaba in Niger, Klaas van Walraven demonstrates how the introduction of borders based on the Westphalian system were used by Sawaba in ways that had never previously been intended, let alone anticipated, by metropolitan France. The border drawn between Mali and Niger was put in place to suit France's administrative whims and bore no relation to the sociocultural and/or political relations on the ground. Van Walraven shows how, in the run-up to independence and afterwards, Sawaba used and transformed the border, which had originally been meant to be an institution of control by the French colonizer, into something that could be used to subvert control.

The way an institutional concept, the 'self-help group' that was introduced by donors in western Kenya, came to be transformed into an independent and sustainable enterprise is the subject of the chapter by Samuel Owuor and Dick Foeken. Their contribution traces the active members of the group and considers how they appropriated the idea of establishing an independent water company to ensure a successful and sustainable supply of cheap and reliable drinking water in one of the informal settlements of Kisumu. Owuor and Foeken sketch the accomplishments of the venture and suggest that similar initiatives could be applied elsewhere on the continent in an effort to ensure the efficient supply of potable water.

The chapter by Marleen Dekker and Bill Kinsey deals with the unintended and unforeseen consequences of land redistribution and resettlement in post-colonial Zimbabwe. Focusing on two large resettlement programmes, they show how these are modelled on an ideal type society with nuclear families that existed in the minds of planners but do not match existing concepts in Shona society. Their contribution demonstrates how a modernist venture engendered by the post-colonial state came into conflict with and was transformed by Shona cultural constraints.

Socio-cultural innovations

Cassava, the food crop from which porridge is made in Africa, is at the centre of Iva Peša's contribution. Initially from South America, cassava was introduced into Africa in the sixteenth century to supplement pre-existing crops and became the staple food crop amongst the Lunda-Ndembu in northwestern Zambia. Peša charts how cassava came to occupy a central role in Lunda society, arguing that it was used as a tool to assert social, economic and group autonomy, whilst also unleashing the area's agricultural potential.

The concept of the white wedding was introduced in the colonial period but it has only been in the last decade that these weddings have truly taken off in Botswana as inflationary institutions of conspicuous consumption. Rijk van Dijk's chapter describes the increased popularity of such weddings and argues that a qualitative distinction needs to be made between adaptation and innovation and the way the wider social environment is affected, involved or changed. Weddings in Botswana mix local customs and engage with a wider world of style, aesthetics, citizenship and class aspirations. Van Dijk argues that it is not the elements of the wedding as such but the skills and competences required by couples to manage their wedding that are truly innovative. The wedding can therefore be seen as a tantalizing 'social cocktail'.

Focusing on a former colonial residence that has been transformed into a tourist guest house on a tea estate and farm, Jan-Bart Gewald describes how innovations, developed and acquired from around the globe, came to be appropriated, transformed and applied in western Uganda. Established by the colonial administrator C.W. Switzer, Isunga Tea Estate drew upon a wide variety of sources from African labour to South Asian agricultural policies as well as Oxbridge norms and values to ensure its success. Gewald argues that any attempt to understand the Isunga Tea Estate will need to take into account these innovations that, although developed elsewhere, have become uniquely Ugandan.

Socio-technical innovations

Automotive mechanical engineers in Eritrea are at the centre of Stefano Bellucci and Massimo Zaccaria's chapter that considers how automobility, which was introduced by the Italians during the colonial era, came to be appropriated and transformed within Eritrea by Eritreans. As an Italian colony, Eritrea experienced a different colonial trajectory and socio-economic development that differed markedly from that in British or French colonial states. Italian rule in Eritrea allowed local Eritreans to work as apprentices to Italian automotive engineers. Bellucci and Zaccaria trace the careers of some of them and show how, under the constraints of ever-changing political rule (Italian, British, Ethiopian and Eritrean), Eritrean mechanics were, through necessity, innovative and consistently succeeded in transforming innovations into something that was their own.

The second contribution by Moses Mwangi and Marcel Rutten focuses on the Arid and Semi-Arid Lands (ASAL) programme in Kenya that is trying to develop solutions that will allow Maasai pastoralists to access water for themselves and their herds in an ecologically sustainable manner. Mwangi and Rutten argue that the 'shallow wells' in Kajiado County were a local innovation, but needed improvement. The authors describe the history of these improved shallow wells, addressing both the technological and institutional aspects and they show that successful water development in semi-arid regions is possible if based on existing knowledge. But the authors also show the successful improvement and sustainability of these shallow wells have been severely undermined by the depletion of the aquifers by large-scale horticultural farmers who are using diesel-electric pumps.

Northern Nigeria came to be connected by rail to the harbours on the Guinea coast in the late 1920s. Although this led to a major transformation in the economy of Northern Nigeria, Shehu Tijjani Yusuf highlights an unforeseen consequence of the building of the line. The railway and all that was associated with it impacted directly on local blacksmithing in Northern Nigeria as it provided raw materials that could be reworked by local smiths, while at the same time it undermined their activities due to imports of cheap manufactured products. Yusuf describes how local blacksmiths used the railway to their own advantage.

Concluding remarks

This aim of this volume has been to show the wide range of innovation and transformation processes in Africa. Contrary to gloomier views of Africa, an alternative discourse on innovation in Africa can be discerned. In 1996, Jane Guyer referred to 'traditions of invention' in Africa and the continent is now slowly but surely being portrayed as one of hope, both in popular and scientific writing (*The Economist*, 3 December 2011; Dietz 2011). It has even been said that Africa 'is buzzing with its own brand of innovation, taking existing technology and adapting it to new uses'.[6] The contributions in this volume provide empirical evidence to support this view and move away from the dominant discourse that sees Africa as lacking innovation.

The different chapters show the diversity, variety and historical depth in innovations that are not necessarily confined to the introduction of a new item, idea or institution but involve the recombination, recycling and/or innovative use of existing objects and ideas. And, last but not least, various contributions highlight the need for further empirical and theoretical exploration of innovation in Africa. To conclude, understanding innovation in Africa needs innovation: innovation of our ideas, perceptions and analytical frameworks on how innovations in Africa take shape and develop. And the volume challenges the reader to embark on this journey.

---

6 'Africa can teach the world to innovate, says author', CNN, 2 August 2011, http://edition.cnn.com/2011/BUSINESS/08/02/dayo.olopade.africa.innovation/index.html

References

Adas, M. (1989) *Machines as the Measure of Men: Science, Technology, and Ideologies of Western Dominance*, Ithaca: Cornell University Press.

Allen, R.E., ed. (1984) *The Oxford Dictionary of Current English*, Oxford: Oxford University Press.

Austen, R.A. & D. Headrick (1983) 'The Role of Technology in the African Past', *African Studies Review* 26(3/4): 163-184.

Bayart, J-F. (1989) *L'état en Afrique: La politique du ventre*, Paris: Fayard.

Bayart, J-F., S.E. Ellis & B. Hibou, eds (1999) *The Criminalisation of the State in Africa*, Oxford: James Currey.

de Bruijn, M.E., F.B. Nyamnjoh & I. Brinkman, eds (2009) *Mobile Phones: The New Talking Drums of Everyday Africa*, Bamenda: Langaa Publishers.

Dietz, A.J. (2011) 'Silverlining Africa: From Images of Doom and Gloom to Glimmers of Hope', Inaugural Address, Leiden University & African Studies Centre, 14 January.

Dutta, S. (2011) *The Global Innovation Index 2011: Accelerating Growth and Development*, Fontainebleau: INSEAD.

Gewald, J-B., S. Luning & K. van Walraven, eds (2009) *The Speed of Change: Motor Vehicles and People in Africa, 1890-2000*, Leiden: Brill.

Goody, J. (1971) *Technology, Tradition, and the State in Africa*, London/Ibadan/Accra: Oxford University Press.

Govindarajan, V. & C. Trimble (2012) *Reverse Innovation: Create Far from Home, Win Everywhere*, Harvard, MA: Harvard Business School Publishing.

Guyer, J.I. (1996) 'Traditions of Invention in Equatorial Africa', *African Studies Review* 39(3): 1-28.

Hall, M. (2010) 'Innovation Africa', Speech hosted by Trustafrica, CODESRIA and United Nations African Institute for Economic Development and Planning, Dakar, 24 August. www.salford.ac.uk/_data/assets/pdf-file/0018/73620/innovation-africa.pdf. Accessed 20 July 2012.

Hodgson, G.M. (2006) 'What Are Institutions?', *Journal of Economic Issues* XL(1): 1-25.

Iliffe, J. (2007) *Africans: The History of a Continent*, Cambridge: Cambridge University Press.

Kline, R. (2000) *Consumers in the Country: Technology and Social Change in Rural America*, Baltimore: Johns Hopkins University Press.

——. (2003) 'Resisting Consumer Technology in Rural America: The Telephone and Electrification'. In: N. Oudshoorn & T.J. Pinch, *How Users Matter: The Co-construction of Users and Technologies*, Cambridge, MA: MIT Press, pp. 51-66.

Kline, R. & T.J. Pinch (1996) 'Taking the Black Box Off Its Wheels: The Social Construction of the Automobile in Rural America', *Technology and Culture* 37: 776-795.

MacCracken, J. (2011) 'Bicycles in Colonial Malawi: A Short History', *The Society of Malawi Journal* 64(1): 1-12.

Macola, G. (2010) 'Reassessing the Significance of Firearms in Central Africa: The Case of North-Western Zambia to the 1920s', *The Journal of African History* 51(3): 301-321.

North, D.C. (1990) *Institutions, Institutional Change and Economic Performance*, Cambridge: Cambridge University Press.

Oudshoorn, N. & T.J. Pinch (2003) *How Users Matter: The Co-construction of Users and Technologies*, Cambridge, MA: MIT Press.

Pinch, T.J. & W.E. Bijker (1984) 'The Social Construction of Facts and Artefacts: Or How the Sociology of Science and the Sociology of Technology Might Benefit Each Other', *Social Studies of Science* 14: 388-441.

Prahalad, C.K. (2012) 'Bottom of the Pyramid as a Source of Breakthrough Innovations', *Journal of Production and Innovation Management* 29(1): 6-12.

Prahalad, C.K. & S.L. Hart (2002) 'The Fortune at the Bottom of the Pyramid', *Strategy and Business* 26: 54-67.

Richards, P. (1985) *Indigenous Agricultural Revolution: Ecology and Food Production in West Africa*, London: Hutchison University Library.

The Economist, 'The Sun Shines Bright', 3 December, pp. 68-70.

Wolf, S. (2007) 'Encouraging Innovation for Productivity Growth in Africa', *ATPC Work in Progress*, No. 54, Economic Commission for Africa.

2

Who killed innovation in the Cape wine industry? The story of a stuck fermentation c. 1930-1986

Paul Nugent[1]

The main constraint facing the Cape wine industry at the turn of the twentieth century was the overproduction of cheap wine from inferior cultivars. In the context of an official commitment to promoting 'good wine', this chapter examines the fate of innovation in the vineyard, the cellar and the marketplace. It argues that innovation in one domain often delivered short-term gains that were checked by the lack of transformation elsewhere. It identifies four key periods of innovation before the mid-1990s: the 1930s when serious attempts to replant vineyards and develop an export market were undermined by the KWV; the 1950s when the discovery of cool fermentation permitted a breakthrough in quality for white wines but drove small producers into the arms of cooperatives that became a barrier to further innovation; the 1960s when, in the context of the de-racialization of the liquor laws, merchant-manufacturers pioneered new brands but failed to conquer the Black market; and the 1970s when independent producers sought to bottle their own (largely red) wine but were constrained by the KWV system of regulation. The KWV protected the small producer and rewarded quantity over quality, largely for political reasons. In the 1980s, overproduction, the smuggling of planting material and the dumping of cheap wine signalled a crisis in the industry that was only resolved following deregulation.

[1] The research for this article was partly funded by a Leverhulme Trust Fellowship and was completed during a fellowship at the Stellenbosch Institute for Advanced Study (STIAS). The author wishes to express his gratitude to both institutions.

Introduction: Tackling the *terroirists*

'Innovation' is usually loaded with positive associations in the arena of science and technology, but it is not as straightforward when it comes to wine. For self-proclaimed *terroirists*[2] fighting a war against the globalization of taste – in which the American wine pundit Robert Parker and his infamous 100-point scorecard are the sworn enemy[3] – 'innovation is almost a dirty word'. Rigaux (2010: 43-46) thus bemoans the manner in which New World producers, such as Australia and California, have flooded the market with brands based on the manipulation of chosen cultivars that are artificially enhanced through irrigation and fertilizers to manufacture certain flavours considered consumer-friendly. For Rigaux, wine that is true to its *terroir* almost literally makes itself and requires minimal human intervention.

Although the *terroirists* enjoy considerable support in French wine circles, the underlying claims do not stand up to close scrutiny. Surely if Benedictine monks came to a sound appreciation of which vines grew best in Burgundy, discovered which parcels of land produced the most complex wines and honed effective methods of vinification, then this was precisely because they were prepared to experiment. Moreover, if each *terroir* – and each *climat* in the case of Burgundy – possesses its own unique attributes, it follows that viticultural and wine-making practices that are suited to one environment will not necessarily be appropriate for another. Hence, a certain amount of irrigation may be necessary in the southern hemisphere to ensure vines survive the dry season, while in the northern hemisphere nature may need a helping hand with the addition of sugar to the unfermented must. The fact that chaptalization is permitted in Bordeaux and Burgundy but is illegal in South Africa and Australia is conveniently ignored.[4] Finally, while nobody would deny that champagne is made from superior grapes, these are blended to create the quintessential manufactured product that sells at a premium in the global marketplace.

---

2 The term is used here to refer to those who argue for the uniqueness of soil and climate in particular locations. The combination, which makes up the *terroir*, is considered the primary determinant of quality and the integrity of the product.

3 The attack on the influence of Parker crystallized in J. Nossiter's film entitled *Mondovino*. See also McCoy (2005).

4 Chaptalization, or the adding of sugar, is resorted to in order to increase the alcoholic strength of wine in regions where limited sunshine results in low sugar levels, and therefore rather thin wines. In the New World, high sugar levels tend to create very alcoholic wines.

Let us therefore accept that innovation has always been at the heart of wine production and focus instead on why it is that some forms of innovation have tended to be more productive than others. This chapter addresses innovation in the Cape wine industry at three levels: in the vineyard, in the cellar and in the marketplace. Wine has been made at the Cape since the mid-seventeenth century, using vines and wine-making techniques derived from Europe, and subsequently rootstocks from the US. Although some of the wine was drunk as table wine, most wine before the twentieth century was fortified and sweet, while much of the total production was distilled to make brandy. What is recounted in this chapter is the shift towards the production and consumption of table wine.

Over the course of the twentieth century, there were bursts of innovation that were stimulated in part by the example of other parts of the wine-making world, but these seldom manifested themselves in the same domain at the same time. Interventions often cancelled each other out or generated merely incremental effects. To better understand the stop-start process of change, it is necessary to track the interplay between science, capital and politics, which has collectively provided the connective tissue linking the three domains. The permutations were complex and the relationship to processes of innovation was far from unilinear. While politics frequently got in the way, it was not the only inhibiting factor. Just as important were misplaced scientific orthodoxies, the reality that individual creativity was often hemmed in by collective responses and a limited consumer market.

Cape wine in an international context (up to 1940)

Across the wine-producing world, the onslaught of phylloxera in the last part of the nineteenth century radically transformed production methods. Having failed to isolate American vines that were both sufficiently resistant to the invasion of *phylloxera vastatrix* (a tiny insect that attacks the roots) as well as being capable of producing a palatable drink, French scientists and farmers scored a breakthrough by grafting European vines onto American rootstocks (Gale 2011). Needless to say, a considerable amount of trial and error accompanied the hard science. Cape wine was ultimately saved by European innovation, although interestingly the Jacquez rootstocks that had proved insufficiently resistant in much of France fared substantially better in South African soils (*Ibid.*: 193). This underlined the fact there were many environmental variables that influenced what worked and what did not.

One of the side-effects of phylloxera and other diseases of the vine, such as oidium, was that Cape farmers became more risk averse. Rather than persisting with cultivars that might be more fickle but that also had the potential to produce better quality wines, the majority of producers – with the partial exception of those in Constantia close to Cape Town where the best sweet wines had historically been produced – opted to replant with varietals that could be relied on to generate generous yields even in adverse conditions. In 1909, around 40% of the country's total production consisted of Groendruif (Semillon), which is used in France for the production of sweet Sauternes as well as for blending in dry white Bordeaux. About half as much came from Steen (Chenin Blanc) and somewhat less from Hermitage (Cinsault).[5] All of these cultivars were capable of producing acceptable quality, but only when yields were deliberately reined in.[6] They could also be heavy bearers and in the South African context were used to make bulk wine that was mostly channelled into the distillation of brandy.

At the turn of the twentieth century, Cape farmers who were teetering on the edge of bankruptcy chose to concentrate on maximizing output. The problem was that when the vast majority of farmers pursued the same strategy, there was a serious oversupply of cheap wine. The export market also dried up as Cape wines acquired a reputation for poor quality, even threatening to extinguish the venerable legacy of Constantia itself. At the same time, the internal market was hemmed in by racially restrictive liquor laws that were championed by an extremely well-networked temperance movement (Nugent 2011). The merchant-manufacturers that bought, blended and then bottled wine under somewhat fanciful labels drove a hard bargain with producers but they also struggled to secure willing buyers. In the redemptive saga of the triumph of scientific and commercial innovation over phylloxera, therefore, we can discern some of the ways in which immediate solutions created unexpected constraints further down the line.

5 Cape Archives (CA) AMPT PUBS C/1/2/1/149, Annexures to the Votes and Proceedings of the House of Assembly, Report of the Commission of Enquiry into the Economic Condition of the Wine Districts (1909).

6 It is a central axiom that vines that 'struggle' produce the greatest quality, while flavour is concentrated when there are relatively few bunches on the vine. Excessive irrigation and inadequate pruning tend therefore to produce thin wine. Modern wine-making involves canopy management (controlling leaf growth) and extensive pruning to reduce the number of bunches. Heavy-bearing wines are inclined to produce a greater number of bunches per vine, other conditions being equal. Shy-bearing varietals produce relatively few bunches and correlate with the so-called noble cultivars.

In the decade after the formation of the Union of South Africa in 1910, wine farmers and merchants were locked in a struggle for survival. For the farmers, the first steps towards resolving the problems of collective action came with the formation of the *Koöperatieve Wynbouwers Vereniging van Suid-Afrika* (the KWV) in 1918. Six years later, the KWV leadership succeeded in persuading Prime Minister Jan Smuts to personally steer through the Wine and Spirits Control Act that conferred regulatory powers on the association. Farmers from Constantia who aspired to produce quality wine were opposed to this intervention on the basis that the KWV was captive to the producers of low-quality wine. At this point, a compromise was struck and wine intended for consumption as wine (so-called 'drinking' or 'good wine') was exempted from KWV control. The producers of ordinary distilling wine were henceforth to be paid minimum prices by merchants, as decided upon by the KWV. The latter also took possession of wine that was considered surplus to the requirements of the trade, paying farmers only for what was considered marketable. The KWV distilled some of this surplus, pioneered alternative products using wine spirit (including perfume and motor fuel) and, when all else failed, resorted to destroying the product.

The KWV never considered its remit as simply one of guaranteeing remunerative returns for producers. Just as important was its mission to persuade farmers to improve their methods in order to build up consumer confidence in a superior product. Following the travails of the late-nineteenth century, overwhelming importance was attached to sound management in the vineyards. In 1928, the KWV poached the country's foremost viticultural expert, Prof. A.I. Perold, from the University of Stellenbosch. Perold's life work was identifying European cultivars that might be suited to South African soils, which also meant paying particular attention to viruses and other afflictions of the vine. He experimented freely, although his most enduring legacy, Pinotage (an unlikely fusion of Pinot Noir and Hermitage), survived by a series of accidents rather than by grand design (May 2009: 60-73). Most of the interaction between research and farming occurred on experimental plots attached to working farms. Alongside Perold, Frank W. Myburgh played a catalytic role as the KWV's wine expert, travelling extensively across the Western Cape to meet farmers and encourage them to improve their wine-making techniques. In this, he enjoyed considerable success.

In the mid-1920s, as the industry began to find its feet once again, there were hopes of a recovery in external demand. In 1923, British imperial preferences were for light wines from the Empire, to which a number of British merchants responded by sourcing supplies from Australia and South

Africa. P.B. Burgoyne & Co. of London actually acquired farming interests in Australia, while Cuthbert Burgoyne, the son of the founder of the company, sponsored a prize for export-quality wine at the Cape Agricultural Show (Faith 2003: 71; Simpson 2011: 229-230). That same year, an ambitious young farmer by the name of H.J. (Manie) Malan[7] was given responsibility for his first harvest on the Alto Farm in Stellenbosch. He bravely invited Perold and Myburgh to sample the wine that resulted and they were so impressed that a sample was sent to Burgoyne & Co., who offered Malan a five-year contract to supply them. Initially, Alto was entirely dedicated to Cinsault grapes but Malan proceeded to plant experimental blocks of Cabernet Sauvignon, Shiraz and Pinot Noir with the advice and assistance of Perold. His premium wines were blends of Hermitage and the 'noble' cultivars. By using a water cooler to permit fermentation at lower temperatures, Malan was somewhat ahead of his time. He was reputedly a stickler for cleanliness in the cellar and consequently his wines travelled better and did not require inoculation with wine spirit.[8] In 1924, Malan won the Burgoyne trophy for 'the best dry red wine approximating to the Burgundies of France' and, after winning the trophy for another three years in succession, was presented with a replica in June 1932.[9]

Malan travelled to Europe in September 1933 as a representative of the Quality Wine Growers' Society in the company of Frank Myburgh and André Simon, a Frenchman who was to gain a reputation as the world's foremost wine expert. Simon took his colleagues on a tour of the most prestigious estates in Germany and France where they were treated to personal tours and tastings by leading figures in the industry.[10] Back in London, Myburgh and Malan set out to make an impression of their own, extolling the virtues of Cape wines to the British press at every available opportunity. Amusingly, Malan was forced to disabuse his audience of the notion that 'native labour was used for foot-pressing of the grapes', proudly asserting that 'to-day we have the most efficient machinery in the world'.[11]

---

7 Manie Malan is profiled in Biermann (1971: 99-102). See Kenney (1981) on Perold who was responsible for a major treatise on viticulture that still commands respect to this day.

8 The use of cool fermentation is mentioned in Malan's recollections in Biermann (1971: 101).

9 *Die Burger*, 3 June 1932.

10 F.W. Myburgh, 'Through the Wine Country of France and Germany', reprinted from *Cape Times*, 30 December 1933, 1 & 4 January 1934.

11 'New Year's Vintage in S. Africa', *Financial News* (undated), clippings in the Frank Myburgh folder in the Distell Archive (Stellenbosch).

He also maintained that South Africa was on the verge of a breakthrough in quality and insisted that its wines were now comparable with any, with the exception of the very finest vintages from Europe. Although boasting was to be expected on what was largely a promotional tour, such confidence surrounding Cape wines was reflected in other contexts too. A case in point is the *Wine Book of South Africa* that was produced for the Empire Exhibition in Johannesburg in 1936. This glossy publication asserted that South Africa was at the forefront of modern methods and carefully balanced claims about the singular advantages of the Cape climate.

Not for the last time, however, the wine revolution stalled. One reason was that Manie Malan was an exceptional case. Rather than planting noble cultivars and ruthlessly restricting yields, most farmers continued to place their faith in heavy-bearing vines. The largest increases in output were recorded in the more northerly districts where irrigation pushed production to the limits. In the ten years after 1924, production doubled with Worcester, Montagu and especially Robertson recording the most substantial increases (see Table 2.1). The fact that farmers were paid for only a part of their output caused some uneasiness amongst members of the KWV. Although the quality producers were not governed by the pricing policies of the KWV, the accumulating surplus had a negative effect on prices across the board. Moreover, the quality producers who were seeking to recreate the Constantia model by exporting sweet wines had grounds for complaint. After 1926, the KWV strategy for dealing with the surplus was to maximize its own exports of fortified wine and brandy in cooperation with a British importer, Vine Products Ltd. The quality farmers needed wine spirit to fortify their wines and it was alleged that the KWV was refusing to sell to non-members. In addition, the association could undercut individual farmers because it did not have to pay for the spirit that it drew from the surplus. The net effect was that the KWV was undermining the financial viability of fledgling export producers, despite professing to be in the business of promoting a transition from the production of bulk wine to drinking wine.

In 1935, just as the *Wine Book* was going to press, evidence of the looming crisis of overproduction led to the appointment of a Commission of Enquiry. The Wine Commission heard evidence from across the industry and issued its landmark report two years later.[12] The case of the quality

[12] Union of South Africa, Report of the Wine Commission (U/G. No. 25, 1937) (Pretoria, 1937).

Table 2.1 Production in leaguers* of wine by district (1924-1935)

District	1924	1934	increase	% increase
Caledon	1,445	2,497	1,052	72.8
Ceres & Tulbagh	3,853	6,284	2,431	63.1
Cape (including Constantia)	3,052	10,537	7,485	245.3
Malmesbury	6,548	10,653	4,105	62.7
Stellenbosch	17,059	27,355	10,926	64.1
Paarl	27,507	42,388	14,881	54.1
Worcester	20,088	48,312	28,224	140.5
Montagu	7,126	19,072	11,946	167.6
Robertson	8,533	34,806	26,273	307.9
Total wine	95,211	202,444	107,233	112.6
Total, including raisins and dried grapes	115,558	226,982	111,424	96.4

* One leaguer = 578 litres.
Source: *Wine Commission* (1937), Table II, p. 12.

producers was presented by the formidable Mrs Albrecht from Constantia who stated in no uncertain terms that overproduction was due to the fact that KWV directors were hostage to the bulk producers who elected them:

> Co-operation as regards wine has been a failure. It has failed to protect quality and has failed to bring stability to the Industry. By the creation of an artificial market a stimulus has been given to the production of inferior wines which has led to an enormous and ever-growing surplus. This surplus is a menace to the Industry and must eventually swamp it. The present position is also leading to the elimination of the highly trained, experienced good wine maker... Before the Ko-operatiewe Wijnbouwers Vereniging started exporting sweet wines, there was in existence growers and merchants who made and exported a high quality article of this type. Due to the unfair monopoly which the Ko-operatiewe Wijnbouwers Vereniging exercises these have been driven off the market. The Ko-operatiewe Wijnbouwers Vereniging is now exporting dry wines and by unfair competition will eventually eliminate the individual exporter.[13]

A written submission from the Quality Wine Growers' Society was equally scathing, noting that a failure to restrict mass production had 'seriously

[13] CA A517 'Wine Commission: Evidence Given before the Commission', Vol. I, 'Memorandum from Mrs A.W. Albrecht', 6 September 1935.

jeopardised the asset of popularity and reputation of our wines'.[14] Although the Wine Commission acknowledged many of the criticisms of the KWV, it concluded that the quality producers would suffer even more from a return to unfettered markets because there would be nothing to inhibit bulk producers from boosting their output even further. The Commission opposed a ban on fresh plantings and argued in favour of creating incentives for the production of better-quality wine from shy-bearing varieties, the responsibility for which should rest with whatever statutory body was created. The report then went on to argue that rather than establishing an independent regulatory authority, it was preferable to build on the existing expertise of the KWV and give the latter a chance to take all aspects of statutory control in hand. In 1940, the KWV was therefore granted control over both good wine and distilling wine, despite the misgivings of quality producers and merchants alike. Whether regulation would entirely stifle innovation remained to be seen but the quality producers had clearly lost an important battle. Although Alto continued to sell some of its wine abroad, the export market was henceforth dominated by products shipped in bulk by the KWV.

The white and the red (1940-1980)

A second burst of creative innovation occurred in the 1950s but then the impetus came from advances in oenology rather than in viticulture. There were two factors that provided the initial stimulus. The first was a two-pronged strategy by the newly empowered KWV to actively encourage farmers to shift to making smaller quantities of 'good wine', whilst simultaneously encouraging consumers to embrace a genuine wine-drinking culture. The aim was finally to resolve the problem of overproduction. There was a belief at this time that South Africa ought to emulate the style of German white wines, which were considered highly aromatic and relatively low in alcohol. To this end, students from Stellenbosch were sent to Geisenheim for further studies, a pattern that continued over the next two decades. In the 1950s, the KWV made a pitch for treating wine as an integral part of a healthy lifestyle when associated with the intake of food. The KWV also took full advantage of the tercentenary of the arrival of Jan van Riebeeck in 1952 and the National Vine Festival to commemorate the plantings by the first Dutch Governor, thereby placing wine at

[14] CA A517, Vol. II, 'Memorandum by Quality Wine Growers' Society'.

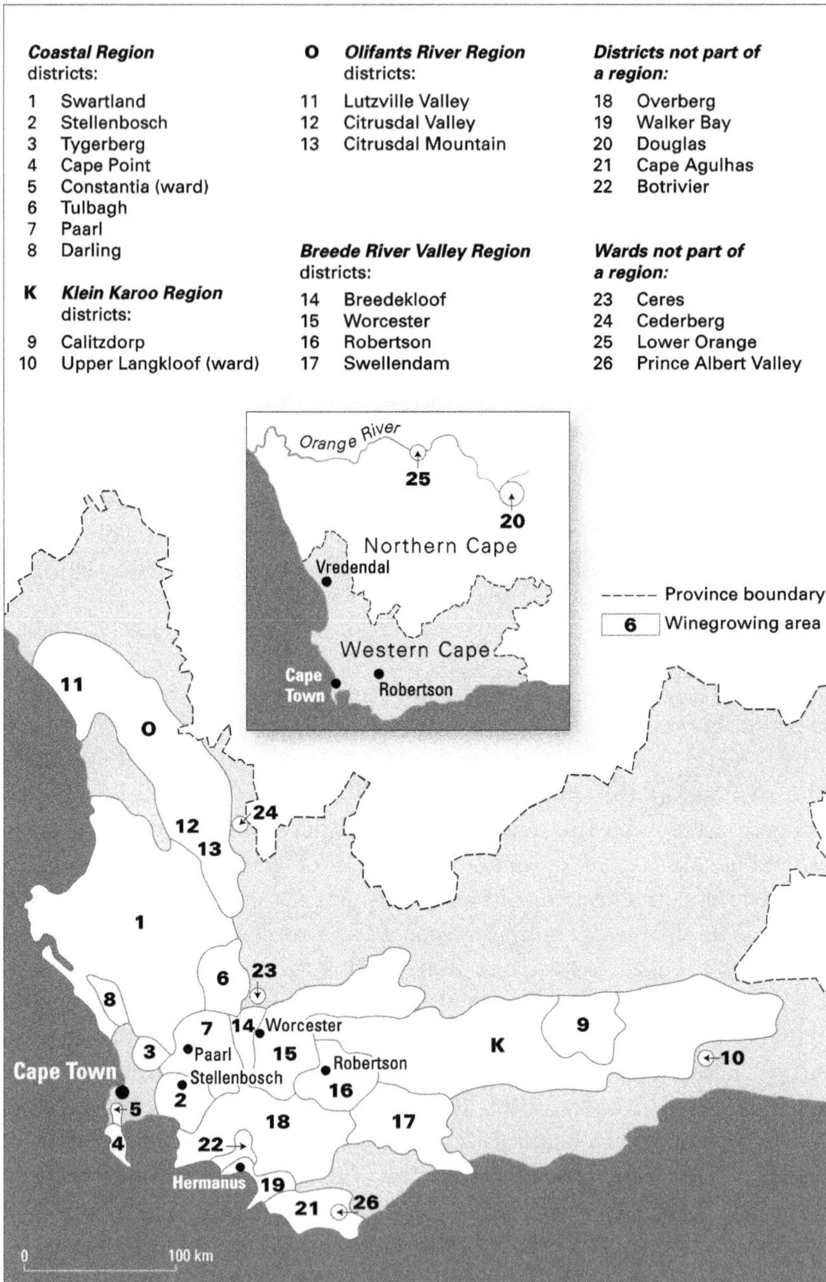

Map 2.1 Winegrowing areas in the Western Cape

the centre of the history of the Afrikaner *volk* (van Zyl 1993: 54-55). In 1952, the KWV also instigated wine festivals in each of the main towns in the winelands, borrowing from a successful Burgundian cultural template (Whalen 2009).

The second factor was the concentration of capital in the liquor industry. In the first half of the century, the ups and downs of the wine trade drove many smaller companies to the wall, with the result being that there were only a few firms left in business by the 1960s. The largest were the Stellenbosch Farmers' Winery (SFW) and Distillers Corporation. These firms were able to assert tighter control over the production process. Distillers established a special relationship with Manie Malan of Alto and A.W. Jordaan of Theuniskraal in 1949 (Venter 2009: 11). SFW records suggest that buyers felt that they were fighting an uphill battle with farmers who often harvested their grapes before they were ripe, either pressed them insufficiently or too vigorously for good wine, and/or treated the fermenting must with insufficient SO_2 with the result that the wines had high levels of volatile acidity. The occasional dead mouse or frog that entered the open tanks did not help! Contact with (hot) air during fermentation was understood to be one of the sources of the many faults. SFW buyers chided and cajoled, and when all else failed they refused to buy contaminated wine. Interestingly, the *Boereboeke*[15] reveal that some of the farmers who went on to cover themselves in glory in the 1970s had been heavily criticized a decade earlier for a lack of attention to basic hygiene in their cellars.

Although it had long been known that excessive heat during fermentation is detrimental to the yeasts and leads to a proliferation of harmful bacteria,[16] the discovery that the esters that impart fragrance to the wine are also lost led to far-reaching changes. There is an on-going debate about who first pioneered cool fermentation in South Africa and the truth is probably that a number of farmers hit on the discovery at about the same time.[17] In the early 1950s, the installation of refrigeration facilities

15 Ledgers that kept a record of the quality and condition of the wine in each farmer's cellar and on the basis of which the SFW made its decisions on how much to buy.

16 A treatise from 1896 by an Edinburgh University professor made this very point and even provided a diagram revealing how a cooling coil was passed through the fermenting vat at Groot Constantia. See Wallace (1896: 140-141).

17 One claimant is the Nederburg Farm, where Johann Graue had been experimenting with cool fermentation in the 1940s. Another is N.C. Krone from Twee Jonge Gezellen in Tulbagh, while a third is what is now Oak Valley Farm in Elgin. Interview with Duimpie Bayly, Stellenbosch, 17 February 2011. The official KWV history accords Graue the credit

and the adoption of steel tanks came to be seen as *de rigeur* for the making of aromatic white wines. N.C. Krone of Tulbagh, who was the first to take the plunge, scooped the main prizes at the Cape Agricultural Show in 1959.[18] His explanation of the science behind cool fermentation and his personal success in following it through, which won the plaudits of André Simon, encouraged other farmers to follow suit.

The transformation in the cellars was accompanied by some aggressive marketing by firms. The SFW and Distillers founded a number of new products, mostly consisting of semi-sweet and slightly sparkling white wines. The brand that broke the mould was Lieberstein. The advertising jingle that became seared into the consciousness of all South Africans went: 'Get together with Lieberstein, any day any time', which signalled that it was a wine for everyday sociability.[19] Initially Lieberstein was sold at the same price as the other SFW brands and was made available in quart, gallon and half-gallon jars. Within a few years it was being targeted at middle-range consumers at a lower price and was sold in cooler units in bars to maximize sales. By 1964, Lieberstein had become the world's largest single brand, which was no mean achievement in a country that had no previous tradition of wine drinking.

The white-wine revolution did, however, generate some downstream effects, not all of which were conducive to innovation in the long term. The cost of steel tanks and refrigeration equipment were beyond the reach of many ordinary farmers who were therefore forced to join one of the cooperative cellars. In the early 1950s, the SFW began to look to the cooperatives to replace some of its contracts with individual farmers. Over the years, a number of these cooperatives cemented their reputation as producers of palatable sweet white wines that won prizes at the annual Cape wine shows. On the positive side, they also encouraged smaller producers to move from producing distilling wine to delivering higher-quality grapes to the cooperative. However, the latter tended to settle into a predictable pattern rather than heading a relentless drive for improvement.

---

for having recognized the principles but notes that it was Krone who introduced the installation of dedicated cooling tanks and other equipment. See van Zyl (1993: 60).

18 Santhagens ran an advertisement in 1959 congratulating Krone for having won eight trophies, including the Burgoyne Trophy, twelve first prizes and one second prize. It slipped in the claim that it was they who had discovered cool fermentation and had made their expertise available to Krone. See 'Gelukwense aan N.C. Krone', *Die Wynboer*, 4 September 1959.

19 Circular sales letter by R.D. Burmeister, 26 September 1961, Distell Archive, Sales Letters, 1951-1962.

The dual emphasis on sound cellar practices and the production of white wines meant that there was, however, much less innovation in the vineyards. From Table 2.2, it can be seen that the share of Semillon had declined considerably since the early twentieth century. The shift towards (white) Palomino grapes, which accounted for 21% of vines, was largely associated with the production of sherry. By contrast, the noble varieties remained conspicuous by their absence amongst the white wines. Chardonnay and Sauvignon Blanc were virtually non-existent, while the so-called Cape Riesling, which was thought to derive from the esteemed Rhine Riesling, later turned out to be the less illustrious Crouchen Blanc. The production of red wines was dominated by Cinsault, much of which continued to be destined for distillation. The limited advances in viticulture reflected an institutional conservatism that ironically derived from the legacy of Perold. The latter had been decidedly enthusiastic about Pinot Noir and even had some success with making wine from experimental farms. But after his retirement, the consensus was that Pinot Noir was unsuited to South African conditions.[20] This assessment reflected the fact that it was trialled in districts that were far too hot for this notoriously 'thin-skinned' grape. Limited quantities of Cabernet Sauvignon, Merlot, Shiraz and Sauvignon Blanc were trialled and released but Chardonnay was considered as unsuitable as Pinot Noir. This largely accounts for the divergent trajectories between the Cape and California, where the adoption of Cabernet, Pinot Noir and Chardonnay, in the context of limited regulation, led to the breakthrough in quality. Most Cape farmers continued to plant the same narrow range of heavy-bearing varieties as before, whereas Californians had access to the best planting material available forced by the narrow price differentials laid down by the KWV. In 1960, the producer of distilling wine received £ 11.1s per leaguer, while the producer of good wine received £ 16.3.6 (KWV 1961: 51). Although the latter might receive a small bonus from the wine merchants, this did not add much to the price. In effect, it still paid farmers to double or treble their output of distilling wine rather than banking on the premium for a smaller volume of drinking wine.

The constraints on further improvement were not immediately obvious in the early 1960s, partly because of the optimism that surrounded

20 In 1958, a researcher from Elsenburg concluded that 'Pinot, the well-known variety used for French Burgundy can not be freely recommended as a result of its poor adaptability to this country'. See De Waal (1958: 32).

Table 2.2 Grape varieties ranked by number of vines over a year old (1959)

Grape	Paarl	W'cester	S'bosch	R'btson	M'mbury	Montagu	Ceres & Tulbagh	Cape	Caledon	Total
Hermitage	11,537,681	10,751,236	14,210,620	6,681,435	4,819,993	1,804,485	1,804,485	1,550,865	1,444,323	53,686,618 27.2%
French grape (Palomino)	5,145,656	6,134,655	1,432,068	11,999,192	8,099,881	6,159,324	1,754,263	459,290	301,730	41,486,059 21.0%
Hanepoot	1,429440	11,353,290	1,018,874	2,635,561	4,000,814	2,374,993	1,045,051	1,054,642	984,752	26,347,417 13.4%
Stein	3,393,264	4,788,147	9,366,796	1,917,645	1,912,591	167,605	843,113	1,488,868	340,108	24,218,137 12.3%
Green grape (Semillon)	8,954,848	2,437,216	1,074,088	152,408	4,623,603	30,300	1,036,011	64,878	546,130	18,919,482 9.6%
Pedro	3,836,935	915,139	1,108,837	51,966	2,765,783	5,400	408,997	109,780	48,082	9,250,919 4.7%
Muscadel	53,740	476,383	22,231	2,864,383	370,929	1,762,802	27,860	62,709	2,100	5,643,137 2.9%
Clairette Blanche	1,608,181	163,956	1,914,755	143,740	295,155	3,500	242,875	79,250	11,120	4,462,532 2.3%
Sultana	7,720	908,660	7	659,291	694,765	295,717	569,288	10,000	-	3,109,448 1.5%
Belies of Kanaan	275,400	751,397	296,609	876,126	354,923	241,796	33,320	49,132	12,706	2,882,409 1.4%
Cape Riesling Crouchen	1,453,051	97,009	677,025	12,200	168,116	2,500	190,037	69,437	3,000	2,672,375 1.4%
Cabernet Sauvignon	340,746	20,051	434,902	17,850	13,500	500	8,000	48,089	2,000	885,638 0.4%

Source: KWV Statistics, *Die Wynboer*, February 1961, p. 12.

the de-racialization of the liquor laws in 1961/1962. While the KWV sought to promote wine as the drink of choice of a supposedly incipient Black middle class, the wine firms sponsored jazz festivals and carried out imaginative promotions in the townships. Although there was an overall increase in wine consumption, African consumers flocked to bottled beer that was vigorously marketed through an association with virility and male conviviality (Mager 2010). The contrasting trajectories between wine and beer were all too apparent to South African Breweries (SAB), which owned the SFW and could easily compare sales of its competing products. The White consumer base for wine remained rather narrow and firms survived largely on the sale of cheap wine to Coloured consumers at the bottom end of the market. More ominously, the wine surplus that was declared by the KWV began to rise again at the end of the 1960s. By this point, consumer boycotts of South African products had led to a serious contraction of the external market.

Despite these underlying problems, it is possible to identify a third period of innovation in the 1970s. This was not driven by the KWV or by firms but instead by individual farmers who saw the important Wine of Origin legislation of 1972 (see below) as an opportunity to target a niche market interested in something other than generic wine. These budding innovators aspired to bottle and market their own products, in which the place of origin, vintage and grape variety would be clearly specified on attractive labels. The farmers who spurred each other on were an unlikely group of friends in Stellenbosch and Paarl: Frans Malan (Simonsig) and Niel Joubert (Spier) were Afrikaners; Spatz Sperling (Delheim) was a German immigrant who had arrived penniless after World War ll; and Sydney and Cyril Back (Backsberg and Fairview) belonged to a small group of Jewish farmers in the Paarl area.[21] What they shared was a desire to plant the noble varietals, especially red grapes, and to improve their wine-making techniques. An especially important innovation was the creation of the wine route in 1971, an idea that Malan had conjured up on a holiday in Burgundy with his wife and the Jouberts two years earlier.[22] Although there were strict regulations as to what could be sold at the farm gate, the wine route enabled participating farmers to make themselves known to ordinary White consumers. As particular wines acquired

---

21 There are some fascinating memoirs and biographies surrounding these pioneers. See Back (1996), Sperling (2005) and van Zyl (2002).

22 Van Zyl (2002: 53).

an almost cult following, other farmers followed suit and new wine routes were delineated.

The merchant-manufacturers initially felt threatened by these developments and applied pressure on farmers to deter them from bottling their own wines. But such was the farmers' desire to be in control of their own products that firms were forced to relent and enable the identity of particular wine farms to be actively promoted. The compromise offered by Distillers was to market wines from specified farms, with the varietal and the producer clearly designated on the label, but as part of their Bergkelder range. This allowed farmers to build up a loyal constituency for their own wines across South Africa, while enabling companies to advance their market share. Over time, Bergkelder acquired a direct stake in some of these farms and took greater control of the maturation process. Dr Julius Laszlo, a Hungarian immigrant, was employed as the cellar master and is credited with pioneering the deployment of French oak barrels. This enabled a breakthrough in quality for red wines that was of equivalent importance to the discovery of cool fermentation for whites in the 1950s. Fridjhon (1992: 48) moreover observed that it was Laszlo who encouraged farmers to draw inspiration from France as the land of red wine, rather than taking Germany as the benchmark. The Bergkelder extended its reach by offering cellaring facilities at their Vinoteque in Stellenbosch. While Bergkelder wines became highly sought after, including the red wines from Alto, the flip side was that producers forfeited some of their independence (*Ibid.*: 19-29).

In the case of the SFW, competition from the independent producers forced the company to up its own game. The SFW had recruited a German winemaker, Günter Brözel, to take over production at the Nederburg Farm in the 1960s. Ten years later, he became responsible for creating some of the leading sweet white and dry red wines that were marketed under the Nederburg label, using grapes drawn from further afield. These various initiatives were assisted by three other developments. The first was the Wine of Origin system, which permitted the designation of 'estates' in the case of quality producers. It was flexible enough to cover distinct farms treated as a unit but sharing a single cellar, as well as wards, districts and regional appellations in which grapes from across a wider area could be used. A tussle between the wine firms and the Estate Wine Producers eventually led to a compromise in which a distinction was drawn between 'estate wine' (such as Kanonkop) and wine that derived in part from an estate (such as Nederburg), each with their own distinct label (van Zyl 2002: 48-49). The second was the proliferation of wine-appreciation societies, including local chapters of the Chevaliers de Tastevins from Burgundy, which tasted and scored wines and prepared the way for local arbiters of taste to emerge across the country. Thirdly, an additional borrowing from Europe was the Nederburg auction, modelled on the annual auctions at the Hospice de Beaune and Kloster Eberbach, which had commenced operations in 1975 (van de Merwe 2000: 11-12). This enabled the SFW and individual estates to showcase and, of course, market their premier wines. While the price bands for quality Cape wines had hitherto been narrow, the auctions now created a price premium for the very best wines that, in turn, provided a spur to further innovation.

The underlying problems were, however, all too apparent. After the Soweto uprising, it was unrealistic to imagine exporting quality wines on any scale, while a highly segmented liquor market meant that there were limits on local demand. In the mid-1970s, when SAB and Distillers fought a bitter liquor war, it is telling that the battleground was beer and not wine. In the peace deal that came into effect in 1979, SAB divested itself of the SFW and was only too happy to assume a monopoly on the sale of beer and later went on to become one of the world's dominant brewing companies.

Contesting regulation (1980-1986)

The fourth period of creative innovation came in the 1990s and saw the first genuine breakthrough based largely on the recovery of an export market. This lies outside the chronological limits of this chapter and would require more extensive treatment than is possible here. I confine myself instead to a brief discussion of the challenges that wine farmers mounted to the KWV system, which is what eventually enabled the industry to reconstruct itself.

For parts of its history, the KWV had led or at least facilitated some innovation. As already seen, it invested in scientific research and after 1940 made a conscious effort to steer farmers and consumers alike in the direction of 'good wine'. But by the late 1970s, many farmers felt that the KWV system was excessively bureaucratic and an impediment to progress. A particular source of grievance was the quota system that was introduced to deal with the problem of overproduction. The quotas, which were not, based on the attributes of the land but on historical production levels, limited how much wine individual farms were permitted to produce. Whereas some farms were allocated relatively generous quotas, others received miserly ones and were only viable if producers resorted to mixed farming. Before 1986, it was not possible to sell quotas separately from farms. In addition, stringent quarantine restrictions prevented farmers from experimenting with different kinds of grapes. Vines that were imported had to pass through a lengthy quarantine procedure and then had to be tested for viruses and general suitability for South African conditions before being certified and released to farmers.

As early as 1973, certain producers began to import vines illegally, a practice that accelerated at the end of the decade as the shift to quality wine intensified. The cultivar that everybody wanted was Chardonnay but many other French and German varietals were also imported illegally. Grafted vines were flown to Swaziland and smuggled overland and, in one case, cuttings were even posted to a recipient at KWV headquarters. As it became public knowledge that the law was being flouted, the government was forced to intervene by creating a one-man Commission of Enquiry, consisting of Justice C.F. Klopper, in 1985. He revealed that the list of culprits was a veritable who's who of the wine industry,[23] including Rhodes Fruit Farms, which was owned by Anglo-American, and Distillers Corporation. Both benefited considerably from the illicit activities of Laszlo. Another Hungarian émigré employed by Distillers, Desiderius Pongracz, was instrumental in disseminating imported planting material, although Klopper found that, on balance, this was probably legal. A leading part was played by respected farmers and winemakers like Nicky Krone (the son of N.C. Krone), Frans Malan, Nico Myburgh (of Meerlust), Sydney Back, Jan 'Boland' Coetzee and Peter Finlayson. At the centre of the web was Danie de Wet, a former student at Geisenheim and considered by many as an Afrikaner 'insider'. He had made his first illegal imports in 1973 and even won a prize for the best Chardonnay, which later turned out to be Auxerrois.

The farmers admitted their guilt, which led Klopper to delve more deeply into the reasons why so many had been driven to break the law. He found that the periods of quarantine had been excessively stringent, which meant that it might take at least ten and possibly as many as twenty years before a new cultivar could legally be planted. He advised against prosecuting producers who had been left with little choice other than to beat the system (*Ibid.*: 30-31). Chardonnay thus made it onto the approved list whereas the Auxerrois that had been planted in error had

23 Report of the Commission of Enquiry into the Cultivar Purity and Illegal Importation of Certain Vine Propagating Material [Klopper Commission], Pretoria (1986: 10-14).

to be stripped out at considerable expense because it was considered an inferior varietal. Many farmers who had been supportive of the KWV or at least regarded it as performing certain useful functions were inclined to blame the conservatism of the association for the affair.

Amongst the farmers who were involved in the Chardonnay scandal was an Oxford-educated advertising executive from Johannesburg who took up wine farming in his retirement. Tim Hamilton-Russell knew something of the wine business, having handled various advertising briefs for the leading wine companies. His ambition was to reproduce Burgundy in Africa and to this end went in search of a suitably cool climate zone to plant his grapes. He found his optimal *terroir* in the Hemel-en-Aarde Valley near Hermanus, which was well beyond the historical borders of the winelands. Remarkably, he was able to acquire a farm with a quota, to which he added a second farm whose position was rather better (Fridjhon 1992: 227-235). He planted grapes on the second farm, claiming that these were for experimental purposes but secretly blended them into the wine from the quota farm. Pulling the wool over the eyes of the KWV inspectors was evidently a source of great entertainment but also meant that he had to disguise what went into his wine on his labels. As a result, he resorted to cryptic numbers and letters that wine lovers were left to decode.[24] Hamilton-Russell experimented with various varietals before finally settling on Chardonnay and Pinot Noir, the classic Burgundian grapes. While De Wet was considered as an insider by the wine establishment, Hamilton-Russell was the complete antithesis. As an English-speaking patrician, he had little sympathy for the KWV and those who spoke for the industry, and as a farmer he was overtly critical of the labour practices of his fellow producers. Having antagonized the wine establishment, Hamilton-Russell fought a lengthy battle over the information that he was permitted to place on his labels. This was still on-going in 1992. It was largely due to Hamilton-Russell that the KWV permitted quotas to be traded separately from farms, which removed one of the shackles on aspiring quality producers. Although many were put out by his methods, Tim Hamilton-Russell acquired a reputation as a visionary for revealing that much of South African wine was actually being produced in areas that were too hot. If there was one individual who was responsible for taking Cape wine to an entirely new level, it was Hamilton-Russell. Today,

24 Interview with Tim Hamilton-Russell, Hermanus, August 2006.

the country's best red wines are made in the Hemel-en-Aarde Valley and in nearby Elgin, exploiting the *terroir* that he first identified.

Conclusion

This chapter has argued that the Cape wine industry underwent bursts of creative innovation that were followed by periods of stagnation and even regression. A pattern that repeated itself was that innovation in one domain produced temporary gains but these were often blocked by a lack of equivalent movement elsewhere or contributed to unforeseen constraints further down the line. In the 1950s, for example, the discovery of cool fermentation enabled wine firms to take advantage of changes in White consumption patterns, but also led to a homogenization of styles for which a resurgent cooperative movement was largely responsible.

In the wine industry, there were many different levels at which innovation could occur or stall. What was planted and where it was planted clearly mattered a great deal, and attention to curbing yields (or not) was at least as important for ensuring quality. Poor cellar practices could spoil the best grapes, although not much could be done to redeem grapes of indifferent quality. Finally, building a loyal consumer base through marketing and advertising fed back to the vineyard and the cellar by creating or limiting financial incentives to pursue strategies based on improving quality. Like France, South Africa resorted to regulation to resolve the recurrent threat of overproduction, the main difference being that it was a cooperative association, the KWV, that assumed devolved functions from the state. While the KWV was formally committed to promoting the shift from bulk wine to drinking wine, many of its own interventions did more harm than good. By virtue of its close association with the National Party (NP) after 1948, the KWV was under pressure to support the struggling Afrikaner farmer (who generally voted for the ruling party) rather than to reward the wealthier innovators who were often English-speaking and/or were considered disloyal. Official pricing meant that there remained a financial incentive for ordinary farmers to maximize output from high-yielding grapes, while aspiring quality producers struggled to secure the additional quotas they needed. When a challenge to the KWV system was finally mounted, it came from a diverse spectrum of producers including prominent Afrikaner farmers, Jews, German, Italian and Hungarian immigrants and recalcitrant English-speakers like Tim Hamilton-Russell. On the whole, it was the outsiders who drove the process of transformation because they had less of a vested interest in conforming to the rules of the game as defined by the KWV and the government.

The KWV was, however, not the only factor in the equation. The wine companies sometimes encouraged farmers to innovate and were especially important in promoting cool fermentation in the 1950s and barrel maturation in the 1980s. But they were also conservative in other respects. The red-wine breakthrough in the 1970s signalled the re-emergence of the independent producer and a deepening appreciation of red wines but the market remained confined to a combination of affluent Whites and poor Coloureds. For a company like the SFW, selling cheap wines to Coloured people continued to sustain the balance sheet, although they were increasingly undercut by cooperatives that were able to dump large volumes of cheap wine in bulk containers by virtue of a legal loophole.[25] Although the SFW blamed the KWV for the crisis in the wine industry in the early 1990s, the companies continued to target the bottom end of the market and were accused of trying to stifle the independent estate producers at the top end.

In the South African context, the pursuit of innovation was therefore continually affected by bureaucratic conservatism, as exemplified by a refusal to embrace Chardonnay and the cumbersome system of plant quarantine, by vested political interests and by basic financial considerations. It was never simply a matter of whether a scientific breakthrough had taken place or innovators could perceive a better way of doing things. This was often the case but the way in which different elements of the overall system combined helped to determine how far a particular innovation would actually travel. In this, there are lessons for the study of innovation more broadly.

References

Back, S. (1996) '70 Years of My Life, 1926-1996', Unpublished Manuscript.

Biermann, B. (1971) *Red Wine in South Africa*, Cape Town: Buren.

de Waal, C.T. (1958) 'Dry Red Table Wines'. In: *South African Wines, Spirits and Liqueurs*, Stellenbosch: Wine, Spirit and Malt, p. 32.

Faith, N. (2003) *Australia's Liquid Gold*, London: Mitchell Beazley.

Fridjhon, M. (1992) *The Penguin Book of South African Wine*, London: Penguin.

Gale, G. (2011) *Dying on the Vine: How Phylloxera Transformed Wine*, Berkeley/Los Angeles: University of California Press.

Kenney, R.U. (1981) *Abraham Izak Perold: Wegwyser van ons Wingerdbou*, Cape Town/Pretoria/Johannesburg: Human & Rousseau.

KWV (1961) *A Survey of Wine-Growing in South Africa 1960-61*, Paarl: KWV.

Mager, A.K. (2010) *Beer, Sociability and Masculinity in South Africa*, Bloomington: Indiana University Press.

[25] This was a recurrent theme is a series of interviews with Lothar Barth, the retired chairman of the SFW, in *Liquor Store Monthly* in 1991.

May, P. (2009) *Pinotage: Behind the Legends of South Africa's Own Wine*, St Albans: Inform and Enlighten.
McCoy, E. (2005) *The Emperor of Wine: The Rise of Robert M. Parker, Jr and the Reign of American Taste*, New York: Ecco.
Nugent, P. (2011) 'The Temperance Movement and Wine Farmers at the Cape: Collective Action, Racial Discourse and Legislative Reform c. 1890-1965', *Journal of African History* 52: 341-363.
Perold, A.J. (1926) *Handboek oor Wynbou*, Stellenbosch: Pro Ecclesia Drukkery.
Rigaux, J. (2010) *Le réveil des terroirs: Défense et illustration des 'climats' de Bourgogne*, Dijon: Darantière.
Simpson, J. (2011) *Creating Wine: The Emergence of a World Industry, 1840-1914*, Princeton/Oxford: Princeton University Press.
Sperling, S. (2005) 'The Memoirs of Spatz Sperling', Unpublished Manuscript.
van der Merwe, R. (2000) *A Magic Blend, SFW, 1925-2000: Its People, Products and Places*, Stellenbosch: Stellenbosch Farmers Winery.
van Zyl, D.J. (1993) *KWV, 1918-1993*, Cape Town: KWV.
——. (2002) 'Frans Malan Van Simonsig: Visionêre Baanbreker en Leier in die Moderne Suid-Afrikaanse Wynbedryf en Landbougemeenskap', Unpublished Manuscript.
Venter, C. (2009) 'History of Die Bergkelder', Unpublished manuscript.
Wallace, R. (1896) *Farming Industries of Cape Colony*, London: P.S. King & Son.
Whalen, P. (2009) '"Insofar as the Ruby Wine Seduces Them": Cultural Strategies for Selling Wine in Inter War Burgundy', *Contemporary European History* 18(1): 67-98.
Wine Book of South Africa (1936) Stellenbosch: Wine and Spirit.

3

Entrepreneurship, colonial monetary economy and the limits of creativity: Appropriating trading stores in Northern Namibia, 1925-1980

Gregor Dobler

Innovations originating from a different social setting rarely remain unchanged when adopted in a new society. Anthropologists have described the creative agency involved in this process and argued against a belief in the homogenizing force of globalized modernity. Innovations come with varying degrees of malleability and not all can be freely adapted to a society's needs, with some innovations engendering fundamental changes when adopted into a new society. This chapter outlines the history of trading stores in colonial Ovamboland, a 'native reserve' and later 'homeland' during segregation and apartheid in Namibia. It shows how, under favourable economic conditions, an outside innovation was quickly adopted by local entrepreneurs who grew to be a new social and economic elite. This choice of development path had consequences that were no longer controllable by the entrepreneurs and that became an economic liability after Namibia's independence. Building on this case study, an attempt is made to differentiate between four types of objects and institutions, whose adoption will have distinct and typical consequences: objects used for their sign value, objects used as tools, tools whose efficiency changes the social context, and objects linked to a set of new practices without which it would be impossible to use them.

Introduction

Innovation comes in many guises. It often emerges within a given social context: through human creativity alone, in inner-societal conflict or cooperation, or in changing social or ecological conditions that demand different solutions to old problems. However when new things become desirable, societies do not always invent them but may import either objects or institutions from others who have already invented them. The old cultural history school in German anthropology, which called this process 'diffusion',[1] was wrong in defining it as the main avenue of cultural change as inner-societal creativity is at least as important as the diffusion of outside inventions, all the more so since processes of diffusion are always linked to, or lead to, innovation. But if seen in its complexity, diffusion is still an interesting variant of social and cultural change and its analysis leads to the core of concepts of continuity and change. What is social about the way we deal with objects, concepts or institutions? How do they change when they are taken over by a different society? What scope is there for social creativity, and how freely can we ascribe meaning to objects and practices? What exactly is the difference between social reproduction and social change?

Such questions form the core of two distinct strands of literature, in social anthropology on the one hand, and history and science and technology studies on the other. Social anthropology is not usually very interested in the invention of things and the history of science. Instead, anthropologists look closely at the receiving end, at appropriation. They see appropriation as the process by which a society actively engages with a new object or institution, fits it into the new context and locally changes its meaning.[2] For social anthropology, innovation is thus the process of social change linked to the active appropriation of a new object, institution or practice through human agency.

Science and technology studies, on the other hand, prefer to concentrate on the spread of innovation and technologies, following an object or institution through its social contexts and analysing the conditions of its acceptance in a society. Innovation has two faces for science and technology studies. Firstly, it describes the invention of a new technique, a new institution or a new object and, secondly, it denotes the process of social adaptation in which the meaning and social role of an innovation is defined. This phase ends with 'closure', the point at which the inno-

1 For an elaborate and comparatively sound propagation of this perspective, see Graebner (1911).

2 For a classic but little-cited example of this trend, see Thurnwald (1932). See also Abu-Lughod (1989), Beck (2009), Burke (1996) and Hahn (2005, 2011).

vation's meaning is defined and it ceases to be new. It becomes part of normality.[3]

The two perspectives are closely linked and describe the same processes but their different use of the concept of innovation can lead to misunderstandings. In the study of science and technology, innovation comes first, before adaptation and appropriation, while for social anthropologists, adaptation and appropriation constitute the innovation. How the object or institution was originally developed is mostly beyond the interest of anthropologists unless it happens to be invented in the society they are studying.

This chapter starts from the second perspective and relates the history of trading stores in the northern Namibian reserve (later homeland) of Ovamboland. This is interpreted as the successful appropriation of an institution developed outside society. The innovation linked to the spread of trading stores appears to consist of new social practices and ways of relating to the world at large that are necessary, although not predefined, concomitants of the integration of trading stores into the local society. What is interesting here are the local social innovations linked to the appropriation of trading stores, not so much the origins of the institution as such or the way it changed its character when spread through diffusion. This chapter demonstrates that the social changes linked to the appropriation of trading stores cannot be understood if the processes of creative appropriation alone are considered.

Anthropological studies of appropriation have often celebrated the creative freedom to adapt externally invented goods, institutions and practices to a society's needs. Globalization, it is argued, brings people into closer contact with the same goods but practices linked to these goods are not prescribed by their original design. 'No imported object, Coca-Cola included, is completely immune from creolization. Indeed one finds that Coke is often attributed meanings and uses within particular cultures that are very different from those imagined by its manufacturer' (Howes 1996: 6).

This perspective has certainly been important as a corrective to all-too-encompassing theories of modernization and globalization. But there are limits to creative adaptation. In this chapter, the appropriation of the trading store is used as a case study to explore such limits. The trading

[3] See, for example, Carroll *et al.* (2003), Haddon (2007), Hess (1994) and Silverstone & Hirsch (1992).

store is a complex institution – an interrelated set of normalized expectations and practices (an institution) that integrates social actors in different lived worlds into a common logic of agency. To understand its meaning, the inherent institutional logics that come with the original innovation have to be taken into account and cannot be entirely discarded when appropriated by a new society. In short, the two perspectives on innovation outlined above have to be combined.

While people were free to adapt the institution to their needs (and made ample use of the opportunity to do so), they were less free to change the wider institutional setting into which it was embedded and which became socially relevant through the appropriation of the trading store, namely consumer capitalism based on wage labour. The link between a useful innovation appealing to many actors and its wider context changed the conditions of local agency. This was not due to any limits in the creativity of appropriation but to the cunning of institutional reason. The last part of the chapter presents a typology of processes of appropriation according to the degree of creative freedom involved in them.

The outside innovation: White-owned trading stores

Ovamboland is the colonial designation of a geographical area in central northern Namibia, an endorheic basin to the north of the Etosha Pan of roughly 400 km × 120 km. Due to higher rainfall that makes the area suitable for subsistence agriculture, it is one of the most densely populated areas in Namibia, and is inhabited today by about a million people. In the nineteenth century, its society was organized into a number of centrally governed polities whose settlement areas were separated by stretches of wilderness but due to an increase in population in the twentieth century, almost the entire area became settled, with people living in scattered homesteads surrounded by fields. In 1906, trying not to disrupt the supply of labour in the context of the Herero wars in the central areas, Germany declared Ovamboland a Native Reserve, made it off-limits to most Europeans and introduced a pass system for contract workers going to the central parts of Namibia. When South Africa took over as the military occupant in 1915 and a mandatory power in 1920, it kept the reserve structure in place and established the first permanent colonial presence in the area. A Native Commissioner was stationed in Ondangwa in a former mission station close to one of the Kings' Courts, and an Assistant Native Commissioner was posted to Oshikango on the border with Angola.

Seen from the colonial capital, the Commissioner's main task was to prevent any disruption in the supply of workers to the colony's mines and farms. Most of the contract workers came from Ovamboland or, via Ovamboland, from Angola. Young men came mostly to look for work and took contracts of about a year, after which (officially) they had to go back onto the reserve. Before going back though, they bought goods in the south, which they could then barter for a profit in the reserve. Most used the wages they had earned from a few years of contract labour to pay for their bride price and set up their own household. The incentive was strong but the process was long, working conditions were harsh and the pay (even multiplied by barter) was low. As in other parts of Southern Africa, the mining economy was short of labour in the early part of the twentieth century, and the colonial administration had no direct means of coercion to induce people to look for work. On the contrary, outside interventions or tax collection were seen as potentially disruptive in the northern areas as long as the colonial presence there was solely built on the co-optation of local rulers.

In this situation, the Chamber of Mines in Lüderitz approached the administration in the early 1920s with the aim of establishing a trading store in Ovamboland. Afraid of exploitation and unrest, which were common around trading stores elsewhere in Southern Africa, the government had so far blocked all attempts by private traders to open a store on the reserve. The Chamber of Mines, however, argued that the store would not be there to make a profit but solely to encourage prospective labour migrants and make trade goods more accessible. If workers could buy trading goods after their journey instead of carrying them home on their backs, the mines argued, their wages would bring them higher benefits. In addition, the presence of modern goods at home, where the families and elders could see them, would increase the demand for money and secure a supply of labour for the mines. This argument was taken up by Native Commissioner Hahn and convinced the Secretary for South-West Africa, who granted the licence.[4]

In January 1925, the first trading store in Ovamboland opened at a place called Ondjodjo, near the Native Affairs Office in Ondangwa. It was operated as a limited company, the Ondonga Trading Company, and was wholly owned by the Chamber of Mines in Lüderitz.[5] Six months later, the

---

[4] See NAN A450 7 2/18, Annual Report 1924; NAN SWAA 1057-135/22 Letter Drew, Native Affairs Windhoek, to NC Hahn, 6.9.1924.

[5] OTC was registered at a statuary meeting on 10 February 1925. The share capital was £ 1000 divided into 1000 shares, of which Hans Härlin, the General Manager of CDM, held 998 and two other CDM directors one each 'in a representative capacity on behalf of the Lüderitz Chamber of Mines', as a letter from OTC explained to the Registrar in 1943. OTC

same company opened a second store in Omafo, in Oukwanyama near the Angolan border. Both stores were named after their locations, 'Ondjodjo' and 'Omafo Store'. They were operated as a monopoly until, to increase competition, a private trader was granted a licence in 1937. Ondjodjo and Omafo stores were taken over by the new South-West African Native Labour Association in 1943, after which the private trader's licence was withdrawn and his store was taken over by SWANLA as well. When the SWANLA was disbanded after the strikes in 1973, the Bantu Investment Corporation in the new 'homeland' of Ovamboland took over the stores.

The history of the white-owned shops is interesting in itself and is closely linked to the history of the institutionalization of contract labour in Namibia, clearly showing how the colonial administration's conflicting aims were played out. But this is not the place to relate their history.[6] Instead we are interested here in the appropriation of the concept of the trading store by local entrepreneurs.

Unlike many other outside innovations, even the white-owned trading stores did not have to be made acceptable by adaptation. They were a huge success from the start. People were enthusiastic about having goods available locally and chiefs from other regions lobbied the administration to have new trading stores opened in their areas. This is unusual in itself, all the more so since the shop was meant to increase labour migration, and Ovamboland society, like many societies, did not always react enthusiastically to outsiders trying to change their priorities. There were two secrets to the stores' success. First, trade as such was not a new concept in northern Namibia. European trade caravans had brought new goods and changed the local power structure during the nineteenth century, providing chiefs with guns and horses and, by introducing new means of accumulation, had encouraged the use of guns in the accumulation of wealth.[7] During the 1920s and later, Portuguese traders from Angola frequently crossed the border illegally into South West Africa. So when Ondjodjo opened as the first permanent store in the area and the first that only sold goods for cash, it offered a more convenient way of trading

was constituted as a Private Limited Company, so no balance sheet had to be submitted (NAN COM 40). The Ondjodjo store opened on 8 January 1925 (NAN NAO SWAA 1056 135/22, Letter NC Hahn to Secretary, 10 May 1938).

6 For a more detailed account of the history of white and locally owned trading stores in Ovamboland, see Dobler (2013).

7 See Dobler (2013), Siiskonen (1990) and Henrichsen (2011).

through a different institutionalization, but it did not introduce an entirely new concept.

Secondly, the shops had a local monopoly. They offered access to goods that had formerly not been regularly available in Ovamboland, so that acquiring them involved either a long and difficult journey or paying a high premium to those who had bought them in central Namibia or Angola. The catch was that shops were only allowed to sell goods for cash. Until World War II, cash was in very short supply in the area as only a few teachers, catechists and government employees were employed in wage labour. The cash used to purchase goods at stores therefore usually involved the wages of migrant labourers. Through their local acceptance, which in itself was linked to a quest for the consumption of new goods, the monopoly stores contributed to the normalization of migrant labour from Ovamboland.

In the long term, the stores crucially influenced the social, political and physical landscape of twentieth-century Ovamboland, becoming important elements in the local social life, models for a new elite economic activity and nodes of urbanization. In the closed reserve economy, they were the first points of contact with a modernizing colonial society, and images of modern consumption were first developed in relation to them.[8]

Local appropriation and the conditions for its success

The four stores kept their monopoly as far as white traders were concerned but after some years, locals started to open their own stores. From the start, this was actively encouraged by the Native Commissioner who saw it as a means of developing the area and promoting labour migration. The first 'native' to apply for a licence to open a store was Simon Galoua from Ombalantu, a Western chieftaincy far from the monopoly stores. He had worked as a police constable in Oranjemund and decided to open a store after returning in 1937. Due to problems with a local headman and some reluctance on the side of the Acting Native Commissioner, his small store

[8] A discussion of the role Ondjodjo played in relation to migrant workers in central Namibia or South Africa is not possible here. From early traders' accounts, it is clear that all of them knew at least the Ondjodjo store, visited it on a regular basis and took their ideas of what a store should look like from these experiences and also had knowledge of other stores, which influenced their concepts.

only opened in 1939 and when obtaining supplies became problematic for all stores during the war, he was forced to close his shop in 1942.[9]

Due to the war, it was ten years before the next wave of locally owned shops appeared and started to change Ovamboland's economic landscape. In 1951, a number of individuals applied for permission to open stores. The first application was made by Teofilus Mafita Tueumuna, a teacher at the Olukonda Finnish mission school in Ondonga. He asked the Native Commissioner to allow his wife, Helena Tueumuna, to open a store in Ondobe, and his brother, Moses Tueumuna, to open a *kafee* (a small eating place) in Oshikango.[10] Over the next three years, thirteen applications were made to the Native Commissioner's Office regarding opening trading stores. The administration needed a year to deal with these applications but finally trading was officially allowed, subject to a licence. In the words of the Chief Native Commissioner: 'It is our policy to encourage Natives to trade in their own areas'.[11]

This was the beginning of a surprising wave of trade entrepreneurship (see Figure 3.1). In the fifteen years between 1956 and 1970, more than 1200 licensed shops were opened by local people in Ovamboland.[12] They were spread fairly evenly across the country (see Map 3.1), with a few clusters developing in the peri-urban centres. Before 1920, reaching a shop had required a journey of one to four weeks. By 1940, people still had to walk or ride for a whole day from most areas to get to one of the monopoly stores but almost every household in Ovamboland had access to a shop within an hour's walk of their home by the end of the 1960s.

Local stores came in different shapes and sizes. The smallest consisted of a shelf in a thatched hut in a homestead, undistinguishable from others and only known to the neighbours by word of mouth, while the largest were huge supermarket chains owned by millionaires. Different as they were, they all had a novel way of organizing the exchange of goods that the monopoly stores had introduced into the region (and migrant workers already knew from their time in central Namibian or South Africa). Before

9 Simon Galoua first opened his store in Ukualuudhi and moved to Ombalantu after problems with Chief Mulua. NAN NAO 25 17/2, Letter NC to CNC, 28 August 1941; NAN SWAA 1056 135/22, Letter Acting CNC to NC, 20 January 1938; NAN NAO 20 11/1 Vol. 11 Annual Report 1938; SWAA 1057 135/22 Telegram NC to CNC, 26 July 1939.

10 NAN NAO 64 17/7, Letter Teofilus Tueumuna to NC Eedes, 3 February 1951.

11 NAN NAO 64 17/7, Letter CNC to NC, 8 August 1952.

12 In 1978, the number of unlicensed shops was officially put at 6075, most of which were '*cuca* shops' selling beer. This, however, was after the sale of beer was officially legalized in 1973 and boosted the number of unlicensed liquor outlets.

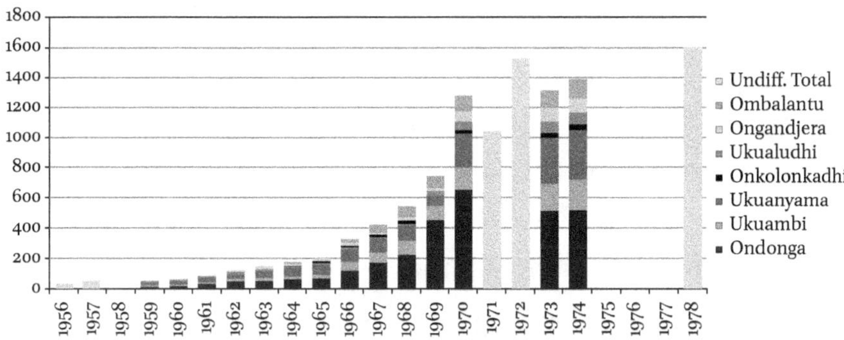

Source: Author's own statistics NAN OVE 11 9/2/5/2, Nominative list of traders.

Figure 3.1 Number of licensed shops in Ovamboland, 1956-1978

the establishment of trading stores, trade had been sporadic, decentralized and embedded in other activities. There are no sources about specialized resident or market traders in the area in pre- colonial times. No 'shops', i.e. fixed venues where private individuals stored goods to sell them on, had existed. Markets had been temporary, mostly developing around important meetings or feasts or when long-distance traders arrived in the area.

Trading stores did not only change the spatial organization of economic exchange, they also introduced a specialized occupation: the career path of the trader. Becoming a trader involved learning a variety of new techniques including bookkeeping, writing applications for a permit, knowledge about a variety of products, and not least the ability to disembed market transactions from everyday reciprocity. It also involved a new degree of regular outside interaction, namely organizing supplies, establishing credit relations and applying for an annual licence renewal. The institutional framework introduced by trading stores thus reached much farther than the few shelves of goods in a local homestead might suggest to the casual observer, and successful traders had to combine local knowledge with knowledge of institutions foreign to most of their customers.

Most traders started small. The first wave of early-1950s traders invested income earned from migrant labour or from one of the few jobs on the reserve (as a catechist, a teacher or a government clerk)[13] but apart from these pioneers, most of the prospective traders gained their first experiences at a very young age. As schoolchildren, they bought sweets in bulk

[13] Government clerks were not allowed to trade while they were in employment, but at least two gave up their jobs and started trading stores.

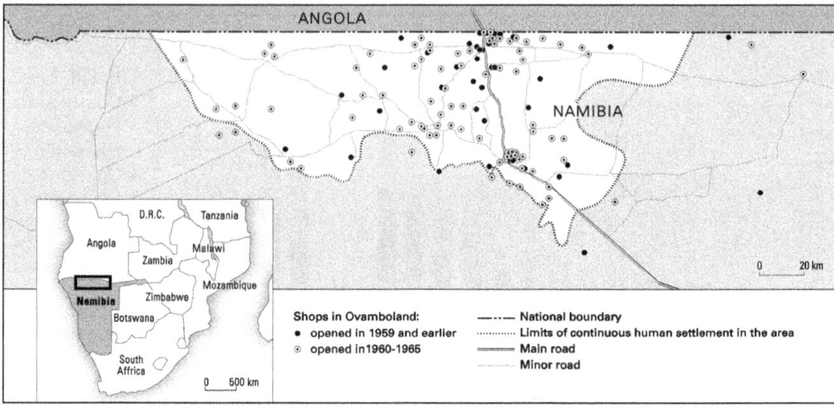

Map 3.1 Shops in Ovamboland opened in 1959 and earlier and 1960-1965

and sold them to classmates, with the more economical soon able to reinvest their profits and move on to slightly more expensive goods like sugar, soap or cooking oil. When they left school and went to the south to work, they did not stop trading but found customers in the labour compounds or the townships. By the time they came back home, some of them had saved enough to invest in a small stock of goods that they brought back home for barter or sale. The risk was lower than the starting capital, since they knew well the goods that were scarce in their villages, and what prices people would be willing to pay if they could avoid having to travel to the next store. People started their businesses as a side-line, trading from a room in their homestead, with most of them only constructing specialized buildings once they had become professional full-time traders.

By the late 1960s, the average, moderately successful trader owned a concrete store with a porch where drinks could be served. These stores were mostly close to the owner's homestead but at the most central place available where different footpaths met or next to a water hole, a local gathering place or the chief's homestead. Entering the store, customers stood in front of a counter that separated the public room from the private space and the goods were stored on shelves behind the counter and handed to the customers by the shopkeeper or an assistant. On the counter were glass cases with smaller goods, scales and weights, and sometimes a cash box. The stores usually served drinks and were often village meeting points.

The shops thus reproduced a function the monopoly stores had gained soon after opening: they became central places in an area formerly shaped by scattered homesteads. No central structures had existed in the villages prior to European influence. The chiefly courts were social centres, and once a person became a chief, his/her homestead grew to accommodate an often sizeable entourage. However they reverted to a normal size after the chief's death and did not result in the formation of permanent centres. The first permanent, larger buildings were the mission stations that were established after the 1870s, followed by the Native Commissioner's Office in today's Ondangwa in the 1920s and, a short time later, the four white-owned shops. As three of the shops were also built close to government posts, mission stations and, in the case of On-djodjo, the native labour compound, the European structures became central places that were functionally differentiated from the remaining parts of the villages. They were the first core areas around which the region's new towns were to develop.

On a much smaller scale, the growing number of specialized village stores embodied a similar functional differentiation. Meeting places had so far either been in somebody's homestead or 'under a tree' in areas designated for public gatherings. The village stores were the first houses exclusively constructed for public use. In this too, they mirrored the white-owned stores. They were often the first 'modern' structures in the vicinity, built not of wood, clay and thatch but of concrete and with roof sheeting. From 1969 onwards, a specialized modern building with a concrete floor and solid walls was no longer optional but was defined as a precondition for a store licence.[14]

Locally owned shops thus relied heavily on the model of 'imported' white-owned stores that were channelled partly (and probably more importantly) through the local appropriation of a Western model and partly due to a favourable administrative policy. Traders and local society, however, modified the institution in important dimensions wherever it did not fit local needs. This concerned changes to size, stock and capital expenditure outlined above and the more flexible transitions between formal and informal businesses. One of the most crucial changes, however, was in the social role shop owners gained locally. Unlike the white managers of the big stores, local shop owners had to make sure they did not appear as greedy outsiders when trading with their neighbours, friends and relatives.

[14] Not all store owners applied for a licence but a surprising number did – mostly because a valid licence was a precondition for buying in bulk at white-owned stores and for applying for a permit to travel to the south to get supplies.

In most peasant societies and frequently in other Southern African reserve economies with white-owned trading stations, traders are seen as leeches that earn money by exploiting their monopoly on the necessary capital to buy and stockpile goods. In colonial Ovamboland, traders seem to have avoided this pitfall by gaining access to elite status and accepting the obligations linked to this role. As owners of new meeting places, they began to play a central role in local political discussions that created the conditions for resistance to colonial rule. Many became community organizers, and quite a number of the early traders left the country to become full-time politicians in exile. Others stayed and took up important functions in the village or local parishes organizing feasts, paying the school fees of protégés, using their cars to provide transport for villagers and generally contributing to the material basis of village social life. Through credit networks that extended all over the homeland, traders functioned as local banks for migrant workers wanting to keep their money in a safe place. Trade was almost the only local alternative to wages earned as a contract worker and traders became the core of this new modernizing elite that was neither oriented towards the local subsistence economy nor towards wage labour in the mines. They found an economic basis in the monetary economy supplied by the wages of migrant workers, and often opposed local headmen linked to a village society that relied on subsistence farming. The increased consumption opportunities the new shops provided shifted the balance of power away from the agricultural elites and towards groups that could command a monetary income. The new trading stores contributed to the conceptual division between the 'homeland', in which families lived and consumed, and the central Namibian towns in which young men went to work. The new villages and towns that emerged around the clusters of shops were the visible expression of this re-centring of the local economy and, in them, novel forms of sociability gained increasing importance. The structural conditions of this appropriation of an outside innovation are traced below and some of the unintended consequences it had for the local society are discussed.

The first precondition of a successful appropriation of the institution of the trading store was the spread of a monetary economy. The shops fed into it without creating it. It had its origins in the mission stations, which paid their employees in cash and sold clothing or Bibles more cheaply for cash than in kind. In migrant labour, cash had the function of being a temporary stash that could be transformed into tradable goods one could take to the village, and the pay of native soldiers during World War II was important in this respect too. These activities brought cash into the area and fuelled demand as soon as local shops offered a way of spending it at home. The years between 1940 and 1950 were the watershed and by 1950, there was enough money on the native reserve of Ovamboland to make locally owned shops profitable. The new shops then offered an incentive to returning migrant workers to bring cash instead of goods back home.

Secondly, anyone who wanted to open a store needed the capital to do so. This rather banal point is rarely analysed in the literature on appropriation but, short of stealing them, one can only appropriate things one can afford to buy (see Beck 1990). The concept may be there for the taking but the actual store is not. And material innovations, like trading stores, often necessitate savings. They change people's aims and thus affect how they frame their own agency. Appropriating an innovation by, for example, buying a car, might involve years of wage labour and might mean not buying a donkey cart, and necessitates acquiring an entirely new set of competences. In the same way, opening a store meant altering one's life choices and neglecting other possible aims.

Necessary investments and changing aims are both clearly visible in the trading stores. In a society in which money was rare, the capital to open a store was not easy to come by. The first local people to open shops in Ovamboland were employees with a regular monthly income, mainly teachers, preachers or nurses working on the mission stations, or returning migrant workers who had put aside several years' wages. From the mid-1950s onwards, when money became more widely used, future traders often started their career by trading small amounts of goods as children, selling units individually and reinvesting the small profit, as already mentioned above. Both strategies involved long-term planning, discipline and foresight but, above all, a re-orientation of one's life plans towards the cash (and often wage) economy. In the 1960s and 1970s, being a farmer and being rich in land and cattle lost its appeal for many young men, to be replaced by the model of a migrant worker who reinvested his wages in trade and became rich through trade.

A third precondition of the success the trading store concept had was the lack of alternative ways to earn a living at home outside subsistence agriculture. Employment was still reserved for the few well-educated teachers, catechists or nurses (who often had their own shops, as well) and all the other men looking for employment had to leave the area as contract workers. Almost no manufacturing took place on the reserve due initially to a lack of money locally to pay for manufactured goods and the comparatively competitive wages in the mining sector. This was later reinforced by the easy availability of industrially produced goods on the reserve. The very presence of a high number of shops in an economy fuelled by outside transfers prevented the emergence of a more sustainable manufacturing industry. Entrepreneurs who had the money and energy to build up a successful company were sure to benefit from trade. Due to the pass laws, which eliminated all outside competition, profit rates remained high and due to the migrant labour system, which geographically decoupled labour markets from consumption, demand remained stable. Stores became even more lucrative when South Africa deployed increasing numbers of soldiers in the area during the guerrilla liberation war that started in 1966. The recruits received wages, a good part of which ended up in the pockets of the owners of local bars and shops.

This led to the fourth precondition of the successful appropriation of shops. The 'homeland' was conceptually perceived as a sphere of agricultural production, consumption and later services, whereas industrial employment and manufacturing took place in the central areas of Namibia. This is where ordinary people went to earn money. At home, they produced food, lived with their families and spent the money they had earned elsewhere. Shops could easily be integrated into this separation of economic spheres but commercial agriculture and manufacturing were far from people's minds. This separation of the world into two spheres shaped by wage labour on the one hand and a combination of subsistence production and consumption on the other is often found in areas with strong outgoing labour migration. It is a crucial, though often overlooked, element in the articulation of domestic and industrial modes of production that have famously been described by Wolpe (1972), Meillassoux (1975) and Evers (1987).

All four structural conditions were necessary for the successful transformation of the trading store and its appropriation into local society. But they did not remain unaffected by the appropriation itself: the presence of trading stores led to feedback effects reinforcing the conditions under which they could blossom. By reinforcing certain trends and weakening others, the appropriation of the trading store had unintended but important structural consequences.

Consequences of the stores' appropriation

Of the many consequences the proliferation of trading stores had, let us first consider the way in which the appropriation of trading stores affected the local economy. As already noted, for those who had saved sufficient capital, opening a trading store was a very promising investment. Running a store allowed the owner to settle down but still earn a regular cash income. The steady flow of outside resources into the region secured demand. Wages for migrant work were relatively low but even low wages provided significant purchasing power in an economy that relied on household farming for everyday subsistence.

For many people in Ovamboland, trade thus provided a way to better their own personal situation. A successful store gave access to a cash income, higher social status and the chance to extend one's social networks, in addition to the comparative freedom that self-employment offers. Trade was attractive as a permanent profession in much the same way migrant labour had been attractive as temporary employment, and it had fewer negative aspects than contract labour. But in both cases, the sum of individual emancipatory decisions set conditions that narrowed the scope of future agency. Once migrant labour and trade had become the normal economic activity for a specific social group, that normality was not only reproduced in social practices and societal structures of relevance but also in concrete institutions, in spatial organization, roads, passports and tax laws.

As long as stores were successful, this did not matter. But stores did not create sustainable economic growth at the local level. Their structural consequences for northern Namibia's economy were negative in the long run, simply because the trade boom that the area experienced in the 1950s and 1960 was nothing but the reverse side of its integration into the global economy through the export of labour power. This stood in the way of other forms of development by absorbing all potential entrepreneurs but did not create the same self-sustainability that small-scale industrial development would have. Owning a shop was hard work but seen from the level of the homeland's integration into the Namibian economy, shops lived on outside rents and on a perversely protected environment.

After independence in 1990, Namibia did away with its homeland boundaries and virtually none of the local shops could compete with mainstream South African, Lebanese or Cape Malay companies that opened branches there. A few richer traders managed to diversify, mostly by investing in real estate but most local shops and supermarkets slowly crumbled. In retrospect, a development path that looked comparatively successful during the homeland era can be seen as having failed and as preventing other, more sustainable forms of development.[15]

---

15 The structural consequences of Namibia's independence in 1990 for local entrepreneurship were similar to those experienced in Senegal and Ghana thirty years earlier. The established local business elite could only benefit from independence in a limited

Appropriation

What does the above show us about the appropriation of outside innovations? To come to conclusions on a more abstract level, it is helpful to take a step back from the stores and look at different types of objects that can be appropriated.

The consequences of the innovation of trading stores were mostly grounded in the institutional framework that came with it, and from which it could not be separated. This is not always the case. Literature on appropriation often looks at material objects that are not as strongly defined by institutions, thus leaving greater scope for social creativity. If I had not looked at stores as such, but at the scales, coins, shelves or counters attached to them, I could have presented much greater differences in the ways different people use them.

Why should this be so? Or, to put it more generally, can we discern typical ways in which different classes of objects shape the relative freedom of social appropriation? As a preliminary suggestion that builds on my earlier work in France (Dobler 2002: 359-369), I would like to differentiate between four types of objects: objects used for their sign value, objects used as tools, objects used as tools with a much higher degree of efficiency and, finally, objects linked to a larger set of institutions.

1. If members of a society take over outside goods and use them for their sign value, they are free to change the practices and ideas connected to them. This may need some clarification. I see the sign value of an object not as opposed to, but as an integral part of its use value, which Marx (1973: 49; 1974: 194) defined as the capacity of an object to satisfy societal

way. They either continued with petty trade or moved their business interests into other sectors, often drawing on connections with the new political class (see Nugent 2004: 63f.). There is at least one important structural difference, however. If we follow Samir Amin's analysis of underdevelopment in Senegal (Amin 1971), well-placed local businesspeople managed to benefit from independence by connecting national markets to former colonial centres in a way that served their own and the centres' interests. Instead of developing into such a 'comprador bourgeoisie', well-placed Namibian businesspeople were able to transform political connections into economic gain by taking shares in Black Economic Empowerment (BEE) deals and by taking over positions in parastatals. This has the same structural effect as a comprador bourgeoisie's share in external resources: it enables the best-placed potential entrepreneurs to reap rents of political connections, not of independent economic activities, and decreases the chances of success for those entrepreneurs who are not part of the elite circles. In Namibia however, BEE policies favoured former, exiled politicians who had the best connections to the new political leaders and were convinced of their right to reap the fruits of decades of political struggle.

needs, whatever their nature. The sign value is that part of the use value that rests on an object's capacity to change the social world: to influence or redefine the social relations between the object's user and other people. Western goods placed on a Mami Wata shrine, African masks put on the wall of an upmarket Amsterdam apartment, refrigerators positioned in savannah homes without electricity – all these appropriations can use objects differently from their original context, thereby leaving a wide scope for social creativity that is only limited in the materiality of the object itself and *local* conventions regarding its meaning.

2. The second type is the appropriation of an object to be used as a tool (by which I understand a use primarily intended to change the material world). This comes in two varieties: there are new tools that serve the same ends (a petroleum lamp replaces a candle) and there are new tools that come with a new practice (the mobile phone enables us to engage in conversations over longer distances). In both cases, the new tools are linked to new practices. We might use electric lighting to shine on the same activities as a candle but the actions involved in lighting it, and generally the corporeal habitus connected to it, differ. Still, if objects are only used as tools embedded in local practices, two societies using the same tools typically differ more than the same society after the innovation differs from its state prior to the innovation.

3. This is no longer true for the third variant. These are tools that, even though they serve the same aim, are so much more efficient than the old ones that they transform social institutions adapted to old tools. The introduction of guns into northern Namibia in the nineteenth century made both war and domination more efficient than they had been when using bows and arrows – so much more so that kingdoms became more centralized, taxation rose and raids became more frequent. This encouraged people to build larger and stronger homesteads and to reconsider the relationship between domination and territory, resulting in changing gender relations in the household and a stronger role for the household head. Such changes were not intended but were logical consequences of a more efficient tool. Automobiles, diesel pumps or machine-drilled wells have had similar consequences.

4. This is all the more pronounced for the fourth type: objects that can only be used in a meaningful way after introducing new practices and institutions that affect the society at large. For example, it is impossible to use a conveyor belt for manufacturing without introducing a division of labour first; and paper money only makes sense if people accept an abstract measure of the value of goods. Of course one can always use

paper money to sniff cocaine, light cigars or decorate one's walls and not every object linked to social institutions in one society has to be linked to the same institutions in another society. But these cases would fall under Types 1 or 2. Marx's (1978, Ch. 2) aphorism that 'the hand-mill gives you society with the feudal lord; the steam-mill, society with the industrial capitalist' uses the wrong example but aptly symbolizes the degree to which some innovations are linked to wider social change. Such institutions frequently develop in stages and an innovation that originates in one society might be adopted into a second society as a new tool, and only there lead to the establishment of a new institution. A third society may then adopt either the original tool or the new institution, with very different consequences.

The trading store is an example of the fourth type of appropriation of an innovation. If you steal a gun and ammunition and have the know-how to shoot, you can use it without any prior social change. Your appropriation of an outside innovation will change your society but will do so within the limits of your use. If you want to use a trading store in a society that has no use for money, you can only do so by changing your society first by introducing new practices. This will set in motion changes with far greater consequences than you intended or were able to foresee.

In the case of the trading store (and this might merit a fifth type), individual agency integrated an ever-wider array of fields of social action into institutions controlled from the outside – or at least connected to the outside and no longer controllable from within. The printing of money, paid jobs, sanitary laws, pass laws, road building and tax regimes are all direct links to social fields beyond local control. Outside innovation was taken over by local actors in a creative, energetic and entrepreneurial way, but through their agency, the conditions for local action were changed in a way that was quickly beyond their control.

This is, of course, what society is all about. Individual actions add up to institutions and to new conditions for agency. When analysing how outside innovations are taken over, it is worth bearing this in mind and not casting one's web too narrowly by concentrating on individual appropriation alone. In appropriation, societies all too often have the same experience as Mephistopheles in Goethe's Faust: 'In the first we are free, in the second slaves to the act' (Scene II).

References

Abu-Lughod, L. (1989) 'Bedouins, Cassettes, and Technologies of Popular Culture', *Middle East Report* 47: 7-11.

Beck, K. (1990) 'Entwicklungshilfe als Beute. Über die lokale Aneignungsweise von Entwicklungsmaßnahmen im Sudan', *Orient* 31(4): 583-601.

——. (2009) 'The Art of Truck Modding on the Nile (Sudan): An Attempt to Trace Creativity'. In: J-B. Gewald, S. Luning & K. van Walraven, eds, *The Speed of Change: Motor Vehicles and People in Africa, 1890-2000,* Leiden: Brill, pp. 151-173.

Burke, T. (1996) *Lifebuoy Men, Lux Women. Commodification, Consumption and Cleanliness in Modern Zimbabwe,* Durham/London: Duke University Press.

Carroll, J., S. Howard, J. Peck & J. Murphy (2003) 'From Adoption to Use: The Process of Appropriating a Mobile Phone', *Australasian Journal of Information Systems* 10(2): 38-48.

Dobler, G. (2002) *Bedürfnisse und der Umgang mit Dingen,* Berlin: Reimer.

——. (2013) *Traders and Trade in Colonial Ovamboland, 1925-1990. Elite Formation and the Politics of Consumption under Indirect Rule and Apartheid,* Basel: Basel Africa Bibliographies.

Evers, H-D. (1987) Substistenzproduktion, Markt und Staat. Der sogenannte Bielefelder Verflechtungsansatz', *Geographische Rundschau* 39: 136-140.

Graebner, F. (1911) *Methode der Ethnologie,* Heidelberg: Winter.

Haddon, L. (2007) 'Roger Silverstone's Legacies: Domestication', *New Media & Society* 9(1): 25-32.

Hahn, H-P. (2005) *Materielle Kultur. Eine Einführung,* Berlin: Reimer.

——. (2011) 'Antinomien kultureller Aneignung', *Zeitschrift für Ethnologie* Special Issue, 136(1): 11-26.

Henrichsen, D. (2011) *Herrschaft und Alltag im vorkolonialen Zentralnamibia. Ein Studie über die Geschichte des Herero- und Damaraland im 19. Jahrhundert,* Basel: Basler Afrika-Bibliographien.

Hess, D. (1994) *Science and Technology in a Multicultural World,* New York: Columbia University Press.

Howes, D. (1996) 'Introduction: Commodities and Cultural Borders'. In: D. Howes, ed., *Cross-cultural Consumption. Global Markets, Local Realities,* London: Routlegde, pp. 1-16.

Marx, K. (1973) *Das Kapital (1), Marx-Engels Werke 23,* Berlin: Dietz.

——. (1974) *Das Kapital (3), Marx-Engels Werke 25,* Berlin: Dietz.

——. (1978) *The Poverty of Philosophy,* Moscow: Progress Publishers.

Meillassoux, C. (1975) *Femmes, Greniers et Capitaux.* Paris: Maspero.

Siiskonen, H. (1990) 'Trade and Socioeconomic Change in Ovamboland, 1850-1906', Helsinki, SHS.

Silverstone, R. & E. Hirsch (1992) 'Introduction'. In: R. Silverstone & E. Hirsch, eds, *Consuming Technologies,* London: Routledge, pp. 1-9.

Thurnwald, R. (1932) 'The Psychology of Acculturation', *American Anthropologist* 34(4): 557-569.

Wolpe, H. (1972) 'Capitalism and Cheap Labour-Power in South Africa: From Segregation to Apartheid', *Economy and Society* (14): 425-456.

4

Frugal innovation in Africa: Tracking Unilever's washing-powder sachets

Cees van Beers, Peter Knorringa & André Leliveld

This chapter explores the phenomenon of frugal innovation in innovation management and multinational enterprises' business models. Frugal innovation is the process of stripping an existing product of unnecessary or luxury attributes to bring it within reach of the four billion people at the Bottom of the Pyramid (BoP) who live on less than US$ 2 a day. It is argued here that this view is too narrow as frugal innovation encompasses not only the transformation of an existing innovation ex ante by the producer who aims to anticipate the specific conditions of the BoP consume, but also the ex post transformation by the consumer who may (un)intentionally change some of the product's attributes while adopting or appropriating it. This two-sided transformation is illustrated by a case study of Unilever's OMO washing-powder sachets in Africa.

Introduction

A description and analysis of how external innovations that have been introduced into African societies are appropriated and transformed in meaning and substance by Africans is the central theme in this volume. Western-based multinational enterprises (MNEs) have played a major role in the introduction and marketing of new products and services in Africa's consumer markets. In many cases, the marketed products and services have not undergone any prior modifications either technologically or in its use. The products and services that were originally designed and produced for consumers in industrialized countries have been exported 'as is' to African markets as well. This implies that the cost-price structure and other product characteristics, such as the use of high-quality ingredients and luxury attributes, have hardly changed and that only middle- and high-income consumers have therefore been able to access the products.

More recently however, a new trend has been observed. MNEs from developed markets have discovered the so-called Bottom of the Pyramid (BoP), which refers to the four billion people worldwide who live on less than US$ 2 a day (Prahalad 2012: 6). To reach these consumers at the BoP, MNEs have become increasingly involved in so-called 'frugal innovation'. This can be seen as a process of stripping existing products of any unnecessary or luxury attributes to bring them within reach of those at the BoP and emerging middle-class consumers in countries such as India, Brazil, China, Indonesia and Russia who are demanding goods and services of global quality but at a local price (Prahalad & Hart 2002). Examples include Unilever's OMO sachets containing small amounts of detergent that are specifically produced for hand washing in cold water; Tata's simplified water purifier that provides poor people with safe drinking water; and low-cost solar lighting (Philips) that is beneficial in places where there is no electricity. But frugal innovation does not only refer to the innovation of the product or service itself. Product and service development for the BoP also require innovation in the business model and the philosophy behind it (Dabke 2011).

In the context of this book, the phenomenon of frugal innovation is an interesting one because it encompasses both the transformation of an existing innovation *ex ante* by the producer, who aims to anticipate the specific conditions of the BoP consumer, and the transformation *ex post* by the consumer who may (un)intentionally change some of the attributes of the product while adopting or appropriating it. Numerous examples[1] of the BoP dominating the development and production of goods have focused on Asia but what is known about frugal innovations in Africa? This would seem to be an area as yet unexplored but one that deserves attention because a large percentage of those at the Bottom of the Pyramid live in Africa and high economic growth rates in the last twenty years in many African countries have led to a rise in the number of middle-class consumers who are increasingly demanding quality products at low prices.

By tracking the history of a well-known frugal innovation for the African market, namely Unilever's OMO washing-powder sachets, and providing

[1] See, for example, *The Economist* (15 April 2010), Govindarajan & Trimble (2012), Nakata (2012) and Singh (2011).

some building blocks for understanding frugal innovation, we are taking a first step into the unknown territory of frugal innovations in Africa. The OMO sachet is an interesting case because it has considerable time depth (having been introduced in the 1980s) and can thus be seen one of the earliest frugal innovations to reach the continent.

The perspective of the multinational (in this case Unilever) that sought to increase its market in non-soap detergents in Africa is adopted here. We acknowledge that this poses limitations on the case study. And although attention is paid to the potential and limits of frugal innovation for local consumers and producers, we were not able within the framework of this explorative paper to conduct fieldwork in Africa among OMO sachet distributors, vendors and consumers, and other competitors in the detergent market. This would undoubtedly have allowed a better description of a perspective from below. Instead, we have had to rely on secondary sources and interview material collected in the Netherlands. We want to emphasize that this chapter is only meant to be explorative and to raise critical issues and questions for a research agenda on (frugal) innovations in Africa.

Frugal innovation: A new innovation pathway?

Frugal innovations are the result of the recent ambitions of mostly MNEs to design and sell products for and to consumers at the Bottom of the Pyramid (BoP), i.e. the four billion people living on US$ 2 a day or less. Nakata & Weidner (2012: 21) mention three reasons for this trend. First, the BoP represents the most significant remaining, or unaddressed, global market. Second, the people at the BoP do, in fact, have financial resources, representing about five trillion US dollars in purchasing power parity, according to estimates by the World Resources Institute (2007). Third, the BoP is receptive and willing to spend money on quality products, provided these are suitable, well-made and reasonably priced.

Reaching the BoP is challenging. The BoP's poverty must be taken into account, along with other issues such as social stigmatization, inadequate housing, infrastructural services and remote locations (Nakata 2012: 3). The BoP is also highly segmented and comprises multiple cultures, ethnicities, literacy, capabilities and needs (Prahalad 2012: 6). The challenge is, therefore, not just simply providing stripped-down versions of products to middle- and high-income consumers, but instead providing value-sensitive innovations that are truly compatible with the unique circumstances of the BoP (Nakata 2012: 3). Such innovations for the BoP are referred to as 'frugal' or 'reverse' innovations.

Frugal innovation does not, however, only refer to the innovation of the product or service itself. It is considered to be a whole new business philosophy that forces organizations to take a fresh look at their business models and innovate in order to come up with quality products at dramatically lower prices (Dabke 2011; Govindarajan & Trimble 2012; Prahalad 2012). Frugal innovation requires business to reconsider and replace existing innovation processes, strategies, finances, partnerships, research methods, business objectives and organizational learning routines (Nakata 2012: 3). Cheap labour does not suffice for this type of innovation as it is more about redesigning products and processes, rethinking the entire production process, discarding unnecessary features and frills, negotiating with suppliers and distributors for the best deals, and finding newer cost-effective means of reaching consumers (Dabke 2011).

In addition, frugal innovation relates to what has been conceptualized as 'polycentric innovation' and designates the global integration of specialized research and development capabilities across multiple regions to create novel solutions that no single region or company could have completely developed on its own (Singh 2011; Radjou 2009). This is in line with two important phenomena that can be observed in innovation processes over the last two decades. First, greater product and technology complexity has increased costs and risks for innovators such that these can barely be dealt with by relying on one firm's own limited resources and capabilities alone. This has pushed companies to collaborate with external partners in developing their innovations. Second, the globalization wave of the last two decades has opened up more possibilities for cross-national alliances that contribute to creating competitive advantage in foreign markets (Lavie & Miller 2008). This gives agency to poor producers and consumers, instead of merely relying on industrial technology from the West.

When developing successful frugal innovations, open knowledge sharing between firms in industrialized countries (which used to be the centre of innovation and technological development) and firms in emerging or developing economies appears to be a prerequisite.[2] Today's fru-

[2] For instance, there has been discussion in India about whether a Silicon-Valley-type phenomenon could be replicated. In this respect, Singh (2011: 79) has observed direct replication would not be feasible but that at the same time there is a trend in India whereby 'multiple industry segments, government, and academia are increasingly joining together with a reverse innovation approach to give a whole new meaning to "GLocal"'.

gal innovations are often the result of an *ex ante* transformative process, mainly informed by local ideas, knowledge and expertise, whereby an existing innovation (a car, a power system, a water filter, etc.) is transformed into something new. This is why frugal innovation is sometimes referred to as 'reverse innovation', referring to the case where an innovation is transformed and adopted first in poor (emerging) economies before 'trickling up' to rich countries (Govindarajan & Ramamurti 2011).

The features of open knowledge sharing between MNEs and smaller firms and being polycentric seem to make frugal innovation a new phenomenon that is not yet adequately covered by innovation theory. Joseph Schumpeter, the father of innovation theory in economics, started by arguing that the innovation and technological change of a nation comes from its entrepreneurs, or wild spirits and that these individuals are the ones who make things work in their country's economy (Schumpeter 1934). But Schumpeter changed his ideas on innovations and firm size in the course of his career, which led to two versions of his work, namely Schumpeter-I and Schumpeter-II (see Breschik *et al.* 2000). Schumpeter-I refers to the idea that smaller (newly established) firms are more flexible and thus more likely to come up with innovations. This is the relationship described above and positions the (small) entrepreneur as the dominant agent in the innovation process in society. Schumpeter-II was formulated later when Schumpeter was convinced that innovation was driven by agents in large organizations as these had sufficient (financial) resources and capital to invest in research and development (R&D). Frugal innovations do not fit neatly into one of these categories. On the one hand, frugal innovation might be driven by individual entrepreneurs, while on the other, it might be driven by MNEs that have the resources to engage in R&D.

Currently, most frugal products are being developed and introduced by MNEs, which would make frugal innovation fit in the Schumpter-II pattern of innovation. But the typical polycentric and knowledge-sharing features and the related business model also have features of a Schumpter-I pattern of innovation and make frugal products and services the result of a kind of hybrid pattern of innovation, which does not fit the classical Schumpeterian pattern.

One could ask whether frugal innovations are really innovations. Frugal innovations tend to be based on reinterpreting, reconfiguring and combining existing practices. Therefore, scholars of innovation with a background in the more technical dimensions of innovation might label these activities as adaptations rather than innovations. We feel, however, that such reconfigurations can be seen as innovations because they can and do provide additional business models to multinational companies as well as new development challenges for (local and transnational) companies in the developing world, with winners and losers. And secondly, these reconfigurations suggest new patterns of innovations that have not been observed before.

Frugal innovation and transformation

Frugal innovation encompasses the idea of the transformation of an existing innovation *ex ante*, i.e. before the product or service is appropriated by the BoP consumer there is a deliberate and organized attempt by producers to reinterpret and reconfigure the product or service to allow for easy adoption and appropriation. Still, like any other product or service innovation, as soon as the frugal product is introduced and marketed to the BoP, unintended transformation (from the perspective of the producer) may occur during the appropriation and adoption of the product by the consumer.

Given that it is mainly MNEs that develop and introduce frugal innovations, also within Africa, one could ask whether, in the context of this volume, something general could be said about whether, to what extent and how African producers and consumers are involved in the *ex ante* transformation in the first place and the *ex post* transformation of a frugal innovation later. The answers to these questions will have relevance for development debates too, as frugal innovation is not only about new products and services but also about new business models with winners and losers. Frugal innovation may be beneficial from an MNE's business perspective but not from an African producer's or consumer's perspective. Is there anything known from existing business and development-oriented literature on these questions?

From a business perspective on research and development, location is a major component of innovation and, therefore, also of frugal innovation. Research and development (R&D) cooperation with different partners originates from different motives. For instance, collaboration with customers is relevant for product innovation (von Hippel 1988, 2005) and process innovations can be stimulated by collaboration with suppliers (Chung & Kim 2003). Cooperation with universities is necessary if basic research is needed (Tether 2002; van Beers *et al.* 2008). Sharing R&D costs, benefiting from the pooling of resources and obtaining assistance in quick market penetration are important motives for cooperating with competitors (Miotti & Sachwald 2003).

A limited number of studies have focused on R&D cooperation with foreign partners (van Beers & Zand 2011). R&D cooperation with foreign suppliers can improve access to new technologies and resources that can stimulate innovation (Eisenhardt & Schoonhoven 1996; Gulati 1999). Collaboration with foreign customers is expected to lead to new product innovations due to the adaptation of products to foreign customers' preferences (Alcácer 2006; Lavie & Miller 2008). These preferences vary in different international markets due to customers' tastes, incomes and/or legal constraints. Collaboration with foreign partners – either firms or customers – has the advantage of getting information from country-specific resources, such as access to knowledge of a specialized workforce, but also information on foreign customers. But we have no examples of how this works for producers and consumers in African countries. Our case study in this chapter suggests that better collaboration with African partners in order to supply goods to the BoP in a profitable way is of great importance. Observing differences in consumption patterns with a Western or Westernized eye is necessary but will not ensure profit and success alone. For example, local African Unilever employees seemed to be less open and connected to the customers in the BoP than had been expected by the company's marketing department.

Von Zedtwitz & Gassmann (2002) report on the MNE location drivers of research and development (see Table 4.1). Location drivers *with regard to research* are mostly related to technology infrastructure and are therefore strongly rooted in existing industrial structures. In general, this is not valid for African collaboration partners and their role is thus very limited. Location drivers *with regard to the development of existing technologies or inventions* are within the sphere of launching the invention onto the market (market access. customer proximity. local market requirements) and within the cost structure of the company (country-specific cost advantages, scaling-up in manufacturing). These drivers can be related to frugal innovations and the BoP, and offer an entrance for African producers.

Table 4.1 is dominated by the perspective of MNEs. For both the profit opportunities in the African BoP of MNEs and the economic and social development relevance, R&D collaboration schemes with African partners need to be analyzed from the perspective of the latter. Is the role of African partners in collaboration with an MNE a subordinate one? If so, how does it contribute to innovation and entrepreneurship activities in Africa? These questions remain largely unanswered in this chapter but should form an important part of any future research agenda.

Table 4.1 Location drivers of research and development

Reasons to locate 'research' in a particular location	Reasons to locate 'development' in a particular location
Proximity to local universities and research parks	Local market requirements
Tapping informal networks	Global customers request local support
Proximity to centres of innovation	Customer proximity and lead users
Limited domestic science base	Cooperation with local partners
Access to local specialists	Market access
Dissipating risks over several research units	Local citizen image
Support of local development projects	Simultaneous product launching
Adhering to local regulations	Use of different time zones
Local patenting issues	Country-specific cost advantages
Subsidies	Facilitating scaling-up in manufacturing
Low acceptance of research in home country	Process innovation to local production
	National protection (intellectual property rights)

Source: von Zedtwitz & Gassmann (2002)

If we look outside Africa, examples can be found of how local producers were involved in transformation *ex ante*. Interest in the popular business press for frugal innovations arose from well-known examples like Hindustan Lever in India. Such anecdotal reports stressed the innovative capacity of local managers of multinational corporations to rethink the marketing strategy of existing products in a way that poorer consumers would also be able and willing to purchase these goods. However, what is much less visible is that the rethinking was based at least partly on tapping into existing business models in the local informal sector. In our view, this is not a coincidence. We expect that the more successful frugal innovations would also systematically draw inspiration from existing local business models and reconfigure these local models in a way that would build a bridge between the MNC and local models. The challenge for future research would therefore be to piece together such trajectories of interactions at the local level and unravel those types of trajectories that are more successful from a business and from a development perspective. For example, if an MNE successfully adopts a local informal business model, this will likely squeeze out at least some of the previously successful informal local producers. The loss of employment in these informal enterprises needs to be included in an assessment of the developmental implications of frugal innovation trajectories.

One central feature to examine in these trajectories is the role of the various actors. Are their interactions more successful when local and international businesses are more equal or unequal in how they contribute to the process? To what extent is genuine cooperation between local and international actors needed for *ex ante* transformation and successful frugal innovation (given that these markets are different and difficult)? In other words, how can MNEs learn more effectively from or interact with local business models to develop more successful frugal innovations? And, from a developmental perspective, in which situations can local businesses more actively participate in the further development and implementation of adapted business models? After all, a vibrant local private sector is a necessary but insufficient condition for local economic development. And, last but not least, where and when does this process lead to the further exclusion of poorer producers, and how can a more explicit focus on inclusive business models be promoted? In short, the interests of the business models aimed at frugal innovation do not by definition coincide. There may be conflicting interests instead.

We hypothesize that frugal innovations with more significant 'local content' and a more prominent role for local entrepreneurs in experimenting with adjusted business models are more likely to be more beneficial to local economic development, both in terms of employment generation and of a more endogenous development path. Moreover, we postulate that a conscious attempt at making more inclusive business models could reduce the crowding out of local entrepreneurs.

Frugal innovation is a process of stripping existing products of unnecessary or luxury attributes to bring it within reach of less affluent consumers. If the consumption of such goods could improve the wellbeing of poorer consumers, this would be an additional and important developmental impact of frugal innovations. However, to develop a more empirically grounded agenda, we need to unpack the issues of both 'stripping' and 'reaching' poorer consumers.

First, stripping products of unnecessary and/or luxury attributes might mean that the first set of attributes to be sacrificed could be environmental and/or labour standards that increase costs but do not necessarily influence the quality of the product. This could result in worsening employment conditions for workers and more polluting production processes at a global level. But such low-road competition is not inevitable. Fresh rethinking and the subsequent re-engineering of a product that satisfies the same consumption needs can also lead to the use, for example, of less polluting or more renewable materials or processes, which in turn could lead to more pro-developmental production processes.

Second, potential consumer markets for MNCs do not consist simply of middle-class OECD consumers on the one hand, and poor Bottom-of-the-Pyramid consumers on the other. While many poorer consumers live in wealthy nations, a rapidly growing segment of middle-class consumers is emerging in the Global South, growing 'twice as fast as the overall world population' (World Bank 2007). This group of consumers, with an annual household income of between US$ 17,000 and US$ 72,000, will increase from 400 million in 2005 to 1 billion in 2030. By 2030, 92% of the global middle classes will be in developing countries (also because many OECD upper-middle-class consumers are labelled as 'rich' because they earn more than US$ 72,000 per household annually). The World Bank (2007: 69) therefore estimates that, by 2030, 'more than a billion people in developing countries will buy cars, engage in international travel, demand world-class products, and require international standards for higher education'.

The question then is how likely it is that the world-class products they will be demanding will include, for example, sustainable and responsible attributes, such as being produced under internationally acceptable labour and environmental standards. This is difficult to predict. On the positive side, we cannot rule out the possibility that the new middle classes might be triggered by sustainable and/or responsible attributes. What we do know about consumer behaviour from OECD countries is that relatively few (around 5%) consumers actually use their 'consumption as voting' (Shaw *et al.* 2005), i.e. employ their consumption pattern to drive companies to adopt sustainable and responsible production patterns. Moreover, there seems to be a major discrepancy between what people say in surveys (around 20% claim to be ethical consumers) and their actual behaviour in a store. An interesting exploratory study by Devinney *et al.* (2005: 15) claims that the gap between consumer commitment to responsibility in principle and in practice is explained by different factors in different institutional settings. Consumers in countries like India and China 'tended to have quite different justifications for their beliefs and behaviour' compared to those in the US or Europe.

In any case, MNCs looking for frugal innovations need to think about three main market segments – the rich, the middle classes and the poor – instead of just two groups as in the past. And at the very bottom end of the consumer market, informal local producers serving poorer local consumers are increasingly being squeezed out of the market by cheap substitute import products from Asia. This further reduces the chance of informal local manufacturers surviving as independent producers. Finally, the dividing lines between the main market segments are becoming increasingly blurred. From a developmental perspective, this leads to questions about the extent to which different consumer demands and budgets lead to different layers of standards of minimally acceptable environmental and social behaviour by producers.

In short, both the production and consumption dimensions of frugal innovations hold potentially important developmental implications that will benefit some and squeeze out others. Having analyzed some considerations relating to frugal innovations, the next section presents a concrete example of a frugal innovation in Africa: Unilever's OMO washing-powder sachet.

Tracking the OMO washing-powder sachet[3]

Any visitor in Africa who has ever been in a local supermarket or village shop will have spotted them – the tiny OMO washing-powder sachets sold for a few cents or shillings. The sachets are a clear example of a frugal innovation meant to serve the Bottom of the Pyramid. At first sight, the sachet is a straightforward product: a small package of washing powder packed in celluloid and bearing the OMO emblem. In Africa, these sachets became known among consumers as Blue OMO because of the blue colour of the detergent. Their introduction was a huge success in Africa. In fact, in many parts of Africa, OMO developed into a so-called generic trademark, or proprietary eponym, to the point that OMO has become the generic name for all non-soap detergents. In recent years, competitors from China and India have been making use of this phenomenon by introducing washing sachets under slightly different names (for example, OEM) and using similar package designs and colours (white, blue and red). The sachet has a rich history with attributes being added to the product either intentionally or unintentionally. This may not be immediately visible but has become an intrinsic part of the product. And these attributes only originate in part from Unilever, as is shown below.

[3] This section draws heavily on an interview with Mr Anton Lenstra, the former Director of the Africa & Middle East Division of Unilever, who was closely involved in the introduction of OMO sachets onto the African market.

Initially Unilever followed its European strategy in the African washing-powder market and simply exported its large packets and tins to Africa. However, these were seldom bought by poorer households as they were much too expensive. The tins were mainly purchased by the rising middle classes and, more importantly for this chapter, by traders and middlemen (and women) operating at local markets. The latter bought large tins and resold the washing powder in small quantities at markets and in local shops. By selling it in smaller quantities, those at the Bottom of the Pyramid could be supplied with high-quality washing powder at affordable prices. The trader realized small profit margins but by being able to sell a large number of small packets out of one tin was able to make more profit than could be realized by selling the whole tin as one unit. Unilever quickly became aware of this practice and reasoned that they could do the same and capture the BoP market from local traders by producing and marketing small sachets of washing powder themselves, thereby regaining some of the profit margin. Unilever still needed local merchants however for distribution and sales but also started to provide supermarkets with the small sachets too. Production of the sachets was also started to prevent reputational damage. After local traders had emptied their OMO tins, they refilled them with locally produced washing powder that did not always meet the minimum set of international standards for washing powder. Complaints were received from customers using this locally produced washing powder (skin infections, itching, etc.). But because the washing powder was sold out of OMO tins, merchants pretended they were selling high-quality washing powder and continued to receive high profit margins on the cheaply produced local washing powder they were selling. This practice had the potential to destroy Unilever's market position in washing powder in Africa; at least this was what Unilever management predicted. This was yet another motivation for product innovation within the washing-powder division.

It soon appeared that African traders were encouraging Unilever to go beyond product innovation alone and it became clear that the product innovation (small sachets of washing powder) also demanded a change in business model. This was necessary to make the BoP market for the sachets competitive with local products, and thus profitable. One change involved Unilever's distribution network, which was too expensive for Africa. Savings were realized and the intention was to set up a business system that would be mainly run by African entrepreneurs. This succeeded in some places but not in others (see Box 4.1) due to a perceived lack of 'business instinct'. In the successful places, substantial savings were made in the

Box 4.1 OMO Distribution

It is morning in the down-at-heel Tanzanian hamlet of Kiwalani, and salesman Sospeter Jackson is busy helping to define the future of marketing to Africa's dollar-a-day economies. Wearing Unilever's battledress (an OMO T-shirt and a yellow 'Key' baseball cap), he has cycled to a tiny outlet beside a dirt road and launched into the daily challenge of bringing his products to some of the poorest people in the world. Salesmen in Unilever uniforms act as mobile advertisements and by travelling door-to-door they develop personal relationships with shopkeepers. They are paid extra if they meet their targets.

'A systematic distribution operation is crucial to the success of any company in a developing economy. But it is a large-scale undertaking', says Rajendra Aneja, MD of Unilever Tanzania. Tanzania has 100,000 retail outlets across the country in 9000 villages. With half the population living below the poverty line, consumers buy rice, maize and flour in tiny quantities every day from mini kiosks in lanes that are too narrow for vehicles. While Unilever delivers goods by van to big shops in towns, it has had to find another form of distribution for outlets in less accessible villages. In December the company came up with a pilot scheme to address this problem: the 'bicycle brigade'.

Salesmen are given bicycles with large boxes welded onto the back to transport small packs of detergent powder, margarine, soap and oil. Each salesman visits about 20 to 30 shops, following a fixed itinerary. Small OMO detergent packs and Blue Band margarine have become market leaders, and Key soap, launched last December, has wrested 15% of a highly competitive market in just eight months.

Source: *Financial Times*, 7 November 2000 ('No Order too Small for Bicycle Brigade that Peddles Products in Rural Tanzania')

distribution network by decentralizing systems of distribution and further cost reductions were realized by downgrading the prod uct, i.e. making changes to the ingredients in the product. Several expensive ingredients were replaced with cheaper ones or were simply omitted.[4] The product was stripped of its flags and banners, a frugal innovation *avant la lettre*. The result was a sachet of washing powder that still had the basic qualities and complied with the health standards of OMO washing powder that was sold in tins, but was now affordable for people earning US$ 1.50 a day. And it still had a minimum cash flow of US$ 0.10 per item. Unilever expected these sachets to open up a huge market for its washing powder

[4] Unilever applied a similar strategy with their Royco soup tubes, which downgraded by removing the more expensive spices while at the same time adding health-promoting ingredients such as minerals and iodine.

at the BoP and that it would become a strong competitor among locally produced washing powders.

The frugal innovation did not work out as expected. The OMO sachets were a success from a sales perspective and turnover increased greatly over figures from the early 1990s due to the introduction of the more affordable economy-sized sachets. But Unilever also expected households to buy at least one sachet a week. Marketing research showed, however, that many households did several months' worth of washing with only one sachet. Further investigation also revealed that households at the BoP did not view the OMO sachets as a product developed for mass consumption. The OMO sachets were used to wash their 'Sunday best' and this powder was only used as a kind of fabric softener, being used in very small quantities in the last hand wash to give clothes the characteristic OMO smell. By doing this, people could show the outside world that they could afford OMO (a status symbol) without letting on that the product was in fact only being used as a window dressing. People did not therefore need to buy new sachets very frequently. In short, OMO was appreciated and bought because of its smell, not because of its intrinsic qualities as a good washing powder. The experiment with the sachet demonstrated to Unilever that for low-income households, OMO washing powder was only bought for its brand name, and that OMO apparently served the same purpose for low-income households that expensive clothes from a certain designer might have for higher income groups. This is a form of conspicuous consumption, i.e. showing others that you can afford a certain brand and thus have a higher social status than those who cannot. And none of the other immaterial attributes of the washing powder, such as its effect on the environment or on the skin (health standards) and the conditions under which it is produced (notably labour conditions) that have become increasingly important in marketing in developed markets, are reasons for African households to explicitly choose OMO sachets. Unilever uses its own minimum standards in these areas for its products but in most African countries there is no government regulation or consumer pressure demanding adherence to any standards.

Experience with the OMO sachets has been a learning exercise for Unilever. Discovering that using OMO is a status symbol in Africa makes it difficult for the company to operate in the BoP market. This has been further complicated by the arrival of cheaper Chinese and Indian washing powders that are currently flooding the African market. The Chinese and Indian products are of a lower quality than Unilever's washing powder but are widely sold at the BoP level because people use large quantities of them before they use that little extra bit of OMO in the final stages of hand washing.

Unilever has acknowledged that it cannot change the reputation and image of OMO washing-powder sachets. And given the size of the BoP and the share it thinks it can capture of it, the company has set its sights on realizing 20% of its turnover from frugal products. At present, Unilever sees frugal products as instrumental in reaching the emerging middle classes in Africa with it more expensive products. The growing middle classes are now eager to buy high-quality products in addition to standard, cheap and often Chinese- and/or Indian-produced products. The OMO sachets are being used to make people who are in the BoP familiar with good-quality washing powder and when they start to climb the social (and wealth) ladder, Unilever hopes they will stay attached to the brand and purchase the larger, and more expensive, OMO tins and packets. This is, after all, where the largest profit margins are to be realized. In Africa, as in Europe and the US, Unilever or its subsidiaries are known for their quality but also for producing relatively expensive products. By introducing cheap sachets in the BoP market, Unilever had hoped to break through this image but the low-income customers, happy to now be able to afford OMO, have maintained the status the frugal product has always had in Africa and are happy to show the outside world that they have moved up the ladder. Who would ever have thought that an OMO sachet would be a vehicle for social mobility when it was first introduced?

Discussion

The story of the OMO sachet shows that introducing frugal innovations in an emerging economy is no easy task. This section attempts to relate the findings regarding the OMO sachet to some of the more theoretical points raised above.

Our first observation concerns the development of the product. Collaboration with customers is considered important in product innovation. Unilever has always done marketing research among its clients in Africa, with local employees visiting households to ascertain consumers' tastes, attitudes and preferences in the field of hygiene products, including soaps and washing powders. But local African Unilever employees have turned out to be less attuned with and connected to their customers in the BoP than was initially expected. Most of the employees were educated people who, given their positions at Unilever, already belonged to the established urban middle classes. They were reported not to be very keen on visiting poor rural households, which meant that feedback to Unilever tended to reflect a middle-class bias and provided less insight into the real needs of poorer people regarding washing powder and its uses. This may partly explain why Unilever did not predict the incidental use of the washing sachets by poor people. It would appear that Unilever collaborated with local business partners in research and development on the product. The company tried though to work through local business structures and partners to make use of their better access to remote markets and their knowledge of local market requirements to realize cost advantages. These drivers to locate research and development near markets are also mentioned by von Zedtwitz & Gassmann (2002) (see Table 4.1). A lack of business instinct among local business partners was mentioned as one of the main reason for failure but it is more likely that the system failed because the uncertainties and variations facing African markets are much higher than in mature markets in industrialized countries and this makes local entrepreneurs in Africa risk averse. A more prominent role by local entrepreneurs in experimenting with adjusted business models would have been beneficial to local economic development, both in terms of employment generation and in terms of a more endogenous development path. But this is still speculative and requires more investigation.

What Unilever did do was to copy a local business model, namely selling small potions of washing powder that poor people could afford. But it wanted to compete with the local business model and did not include local business partners. Although the sachets have been quite successful in terms of sales, turnover and profit, they have not fully succeeded in outperforming local producers and traders. Poor people still use local varieties and OMO sachets are considered 'luxury' products that are only used for specific purposes and 'Sunday best'. Local producers and consumers are not only passive recipients of innovations but are agents that can be innovative and influential.

As mentioned earlier, frugal innovations have brought expensive products within the reach of poorer consumers. If consumption of such goods improved the wellbeing of less well-off consumers, this would mean that frugal innovations could have an additional and important impact. Stripping products of any unnecessary and/or luxury attributes could, for example, mean that the first set of attributes to be sacrificed might be standards that increase costs but do not necessarily influence the physical quality of the product. This could easily result in environmental, health or labour standards being sacrificed. In the case of washing-powder sachets, Unilever did remove the luxury attributes, namely the more expensive ingredients, but did not lower its own standards for health safety, i.e. that washing powder should not result in skin irritations or diseases. As elsewhere in the developing world, customers in Africa rarely demand minimum standards themselves. For Unilever though, maintaining its own standards is a matter of not losing its good reputation globally, producing quality products and keeping its customers in mature markets, who are increasingly concerned about issues of sustainability, satisfied. Whether the real BoP can be reached remains to be seen because over the last two decades emerging players in Africa, like China and India, and local producers too have had no qualms about lowering standards below minimum health and safety standards, which allows them to sell their products at rock-bottom prices, thus capturing the majority of those at the BoP. Unilever has thus been 'sandwiched'. At the top are the consumers (organizations) in the mature markets in Western countries that are increasingly concerned about issues of sustainability in developed and developing countries and who do not accept a further lowering of standards. On the other hand, local producers and new international players are stretching minimum standards to the limits, and sometimes beyond, and are thus able to capture a large share of the market. Local producers and interests have pervaded Unilever's intentions.

The case described above also shows that the BoP is a segmented market and not a homogeneous group of 'the poor'. Producers of frugal innovations cannot assume that their innovations will serve the whole of the BoP. This is not only true for Nano automobile producers, who will never reach the real bottom of the pyramid, but is equally valid for washing sachets. Within the BoP, there are still many people who cannot afford to buy the sachets or who can only afford to use them occasionally, while those in higher income groups may prefer tins of washing powder or even drift away from quality washing powder altogether because washing powder is, in their social strata, no longer a status symbol. So they are purchasing cheaper washing powders again, a trend that can also be seen in mature markets in industrialized countries.

To return to the main topic of this volume, namely transforming innovation in Africa. This chapter has considered whether, by addressing frugal innovation for Africa, product innovations brought into Africa by foreign parties can increasingly become influenced and transformed *a priori* by Africans, before the products have entered the African market. The case of OMO washing sachets could indicate that this trend is still embryonic in Africa. The sachet is certainly a frugal *product*, a serious attempt to take account *a priori* of the preferences of the BoP, but the innovation and business model underlying the product has proven to be frugal to a certain degree only. Further research into processes of innovation, the players in Africa and a wider range of products is needed. Will African countries develop into places where polycentric innovation centres will develop frugal products that in turn will be upgraded to products that may well be able to serve developed markets too? And where will the agency behind innovation place then? With international or local producers, with local consumers, with international or local distributors? Whether African countries will proceed in this direction is an open and intriguing question that the authors of this chapter hope to address in future research.

References

Alcácer, J. (2006) 'Location Choices Across the Value Chain: How Activity and Capability Influence Collocation', *Management Science* 52(10): 1457-1471.

Breschi, S., F. Malerba & L. Orsenigo (2000) 'Technological Regimes and Schumpeterian Patterns of Innovation', *Economic Journal* 110(April): 388-410.

Chung, S. & G.M. Kim (2003) 'Performance Effects of Partnership between Manufacturers and Suppliers for New Product Development: The Supplier's Standpoint', *Research Policy* 32: 587-603.

Dabke, S. (2011) 'Emerging Markets – The Epicentre for Creativity and Innovation': http://www.docstoc.com/docs/82388590/Emerging-marketsThe-Epicentre-for-Creativity-and-Innovation. Accessed 4 September 2012.

Eisenhardt, K.M. & C.B. Schoonhoven (1996) 'Resource-Based View of Strategic Alliance Information: Strategic and Social Effects in Entrepreneurial Firms', *Organizational Science* 7(2): 136-150.

Govindarajan, V. & R. Ramamutri (2011) 'Reverse Innovation, Emerging Markets and Global Strategy', *Global Strategy Journal* 1: 191-205.

Govindarajan, V. & C. Trimble (2012) *Reverse Innovation; Create Far From Home, Win Everywhere*, Harvard, MA: Harvard Business School Publishing.

Gulati, R.D. (1999) 'Network Location and Learning: The Influence of Network Resources and Firms Capabilities on Alliance Formation', *Strategic Management Journal* 20(5): 397-420.

Lavie, D. & S.R. Miller (2008) 'Alliance Portfolio Internationalization and Firm Performance', *Organizational Science* 19(4): 623-646.

Miotti, L. & F. Sachwald (2003) 'Cooperative R&D: Why and With Whom? An Integrated Framework of Analysis', *Research Policy* 32: 1481-1499.

Nakata, C. (2012) 'From the Special Issue Editor: Creating New Products and Services For and With the Base of the Pyramid', *Journal of Production Innovation Management* 29(1): 3-5.

Nakata, C. & K. Weidner (2012) 'Enhancing New Product Adoption at the Base of the Pyramid: A Contextualized Model', *Journal of Production Innovation Management* 29(1): 21-32.

Prahalad, C.K. (2012) 'Bottom of the Pyramid as a Source of Breakthrough Innovations', *Journal of Production and Innovation Management* 29(1): 6-12.

Prahalad, C.K. & S.L. Hart (2002) 'The Fortune at the Bottom of the Pyramid', *Strategy and Business* 26: 54-67.

Radjou, N. (2009) 'Managing the New Trajectory of Global Innovation', Centre for India and Global Business, Cambridge Judge Business School, Cambridge, UK. http://www.cfr.org/content/meetings/New_Global_Trajectory.pdf. Accessed 4 September 2012.

Schumpeter, J.A. (1934) *The Theory of Economic Development: An Inquiry into Profits, Capital, Credit, Interest and the Business Cycle*, Cambridge: Harvard University Press. Translated from *Theorie der wirtschaftlichen Entwicklung* (1911).

Singh, M.G. (2011) 'Innovation in India: Affordable Innovations'. In: Dutta, S. *The Global Innovation Index 2011, Accelerating Growth and Development*, Geneva: INSEAD, pp. 77-86.

Tether, B. (2002) 'Who Cooperates for Innovation and Why: An Empirical Analysis, *Research Policy* 31: 947-967.

The Economist (2010) 'First Break All the Rules. The Charms of Frugal Innovations', 15 April.

van Beers, C., E. Berghaell & T. Poot (2008) 'R&D Internationalization, R&D Collaboration and Public Knowledge Institutions in Small Economies: Evidence from Finland and the Netherlands', *Research Policy* 37(2): 294-308.

van Beers, C. & F. Zand (2013) 'R&D Cooperation, Partner Diversity and Innovation Performance: An Empirical Analysis', *Journal of Innovation Product Management*.

von Hippel, E. (1988) *The Sources of Innovation*, New York: Oxford University Press.

—— (2005) *Democratizing Innovation*, Cambridge, MA: MIT Press.

von Zedtwitz, M. & O. Gassmann (2002) 'Market versus Technology Drive in R&D Internationalization: Four Different Patterns of Managing Research and Development', *Research Policy* 31: 569-588.

World Resources Institute (2007) *The Next 4 Billion*, Washington, DC: World Resources Institute.

5

Mobile cash for nomadic livestock keepers: The impact of the mobile phone money innovation (M-Pesa) on Maasai pastoralists in Kenya

Marcel Rutten & Moses Mwangi

Kenya has become the world's market leader in mobile money since 2007 and today one in three mobile money users in the world is a Kenyan. This chapter discusses the application of mobile money (M-Pesa) by the Maasai pastoralists of Kajiado County in southern Kenya, shedding light on the impact of a mobile-phone-based money system on the local economy. For a small fee, money can be uploaded through a network of M-Pesa agents and sent to people living in faraway corners of the country or to service bills. The technology has proved to be a way of overcoming some of the obstacles that prevent the rural population from accessing financial services. The mobile money system has changed the way people manage their money and organize their livelihoods. In the Kenyan rural pastoralists' set-up, there are many people who have mobile phones but no bank account and for them mobile money is an extremely attractive proposition. Issues of monetary gain, time advantages, convenience, security, insurance during periods of stress and other innovative developments are discussed in this chapter.

Introduction

Since 2007, Kenya has been leading the way with the M-Pesa mobile-phone money transfer service that has transformed the lives of millions of people and businesses in the country. As a result of its rapid growth,

Photo 5.1 Mobile cash [Photo: *Safaricom*]

the application has received significant attention and the diffusion of M-Pesa in the Kenyan economy has been profound. The new technology, a venture offered by Kenya's Safaricom and Vodafone, is the first system to operate on a large scale. It has revolutionized cash flows in social and economic spheres and has changed unsafe, slow and occasional barter trade into a safe, fast and 24/7 mobile-cash-based trade, fundamentally changing the management of the local economy. This has led to greater ease in the way money can move around, not just between individual consumers but among small and medium-sized enterprises as well. The package has been particularly useful in the rural areas where conventional banks are not widespread.

While M-Pesa was seen early on as a no-frills, if somewhat hi-tech, product that was aimed at society's unbanked, it was initially adopted more widely by the better-off. However, the service has ultimately spread to households that are poorer, less well-connected to the financial system and located in more remote areas. While the representation of all segments of income distribution in users' profiles has grown, proportional growth has been highest among those at the bottom of the scale. The Maasai pastoralists who occupy the remote southern Kenyan rangelands are a good example and epitomize a group of innovative technology users that have recently seen their money transfer options revolutionized as a result of the availability of mobile money technology. This technology has provided suitable e-commerce infrastructure solutions in their business processes. To understand the impact of M-Pesa on the socioeconomic position of such people and the possible wider cultural implications, interviews were conducted among three groups of respondents: livestock traders, youth and women. This chapter focuses on the impact that the M-Pesa innovation has had on the first group, namely livestock traders.

The mobile-phone innovation in Kenya

The unprecedented spread of mobile devices, wireless networks and mobile communication services across Kenya has resulted in efficient communication, payments and marketing systems for an increasing number of people. There were more than 28 million mobile-phone users in the country by the end of 2011 compared to fewer than 15,000 in 1999 (CCK 2012). Mobile-phone penetration had thus reached 71.3 people per 100 inhabitants, an increase of 7.9% compared with the year before (CCK 2012). The challenge is now to roll out the network to the rural areas, especially the arid and semi-arid lands. Nearly 68% of Kenyans live in remote, rural areas with just 32.3% living in urban areas (KNBS 2010).

Kenya has a low fixed-telephone penetration rate with only 283,546 fixed lines, a figure that has fallen by 20% in the last year (CCK 2012). Most fixed-line subscribers (90%) are urban residents. The decline in fixed lines has been caused by an increase in mobile subscribers as well as the continuing theft of copper cables. In fact, the decrease was almost 60% but fixed wireless, a technique introduced in 2006, has offset some of the losses and is now the main fixed-line technology. A substantial proportion of subscribers have mobile phones as their only means of communication (Vodafone 2005), confirming that, as elsewhere in Africa, mobile phones are being used as a substitute for, rather than to complement, a landline (Donner 2007). A mobile phone is seen to offer advantages over fixed lines that, in addition to their inertness, often break down. Questionable bills are also an issue.

Analysis of mobile-device-based payments in Africa has previously indicated that the use of mobile payments is conditioned by non-market factors related to financial and technical literacy (Duncombe 2009). However surveys in the same period indicated that M-Pesa was reaching the majority of Kenya's poor, unbanked and rural populations (Suri *et al.* 2010). This implies that the use of mobile money transfer services in Kenya defies the arguments made by Duncombe & Boateng (2009) that the overall level and pace of adoption of m-finance services in developing countries is relatively low and confined to more affluent users. Most of Kenya's poor and unbanked have embraced the use of mobile money transfer services to store and make payments because the services offer cheaper and more secure alternatives to the existing informal money transfer channels and everyday transactions.

Box 5.1 M-Pesa innovation

Launched in March 2007, the M-Pesa innovation is perhaps the most successful service of its kind in the world. Originally funded in part by a grant from the UK government, it was supposed to be a micro-finance loan repayment system that enabled micro-finance institutions to offer more competitive loan rates to their users, with reduced costs for dealing in cash. The users of the service would gain by being able to track their finances more easily. After running a pilot scheme for about nine months, the system showed that customer behaviour suggested that it was actually much more useful as a person-to-person money transfer system. Customers adopted the innovation and transformed it for a variety of alternative uses with different value propositions. The service accommodates all types of needs. Customers can send up to KSh 140,000 a day with a maximum of KSh 70,000 per transaction.[1] Account balances of up to KSh 100,000 are accommodated by M-Pesa accounts and low demands can be serviced with transactions of as little as KSh 10. This accelerates the movement of cash, particularly among low-income earners. Sending costs to other registered M-Pesa holders range from KSh 3 to KSh 100 for sums from KSh 10-50 up to KSh 50,000-70,000. Withdrawing or depositing money at an M-Pesa agent will cost KSh 10 to KSh 300 for these amounts, respectively. M-Pesa services have led to increased ease in moving money around. In addition to sending home remittances across the country and making payments, M-Pesa customers can deposit and withdraw money from a network of agents that include airtime sellers and retail outlets that act as banking agents. M-Pesa has become a 'mobile wallet', especially for the huge proportion of consumers that do not have a bank account. In fact, the service does not require users to have a bank account as such, which is an important point in a country where millions of people remain unbanked. It allows account holders to purchase electronic funds at M-Pesa agent outlets and send the electronic value to any other mobile-phone user who can then redeem it for conventional cash at an M-Pesa outlet. To transact, you need to deposit money in your M-Pesa account at an outlet where the agent will make the transaction on the customer's behalf using their handset. To put money into an M-Pesa account it is necessary to go to an authorized M-Pesa agent

>>>

[1] At the end of 2011, € 1 was worth KSh 110.

Box 5.1, ctd.

with one's phone and an ID card. The agent will send the person e-money in exchange for cash. To withdraw cash, a person has to give their phone number to an agent and show their ID card. Using one's own phone's 'withdraw cash' menu allows a person to enter the agent's number, the amount wanted and a personal identification number (PIN). After confirmation, the agent will give the person the cash and request a signature as proof of receipt.

The growth in mobile-phone ownership has been instrumental in the development and usage of M-Pesa and two other money transfer services. In the year to 30 June 2011, 364 million mobile-phone money transactions were recorded, an increase of almost 45% over the previous year. The customer base of money transfers grew even more from 10.4 m to 18 m customers. The value of the transactions transferred increased by almost 54% from KSh 597 bn to KSh 919 bn. In June 2008, the figure stood at only KSh 61 bn. M-Pesa accounted for 64.4%, Zap 24.4% and Yu 11.2% of all transactions. A fourth operator (Tangaza) was also granted permission and started mobile banking operations in January 2011. (CBK 2011: 55-56).

The conversion of cash into electronic value (and vice versa) is performed at M-Pesa agency points that are paid for by exchanging these two forms of liquidity on behalf of customers. These are also the places that customers turn to for help, and not Safaricom. The number of agents has grown fifteen-fold from 3011 (30 June 2008) to 46,588 (30 June 2011), which underlines the successful diffusion of the service. The number of ATMs has also grown, but only from 1510 (2008) to 2183 (2011) and from a value of KSh 396 bn to KSh 623 bn over the same period (CBK 2011: 55). With a subscriber base of over 18 m, Safaricom provides mobile and fixed voice and data services on a variety of platforms. At the end of 2011, M-Pesa peaked at 285 transactions per second from 154 in 2010 for the same period, an increase of 85% (DN 2011). M-Pesa, with its 14.9 m customers, witnessed an all-time high of transactions going through the system over the end-of-year festive season as subscribers sent money to various destinations. This is further proof that it is East Africa's largest financial service provider.

M-Pesa's transformation of the finance sector

Safaricom has enhanced the M-Pesa product by providing new functionalities and services to customers. These include Pay Bill/Customer to Business Payments that allow M-Pesa customers to pay their bills to partner organizations via the M-Pesa menu. This functionality has notably also been used to collect customer donations for a number of charities. The other service is Business to Customer Payments, which makes it possible for organizations to disburse cash to large groups of people. The service is particularly useful for companies that pay salaries or wages to employees

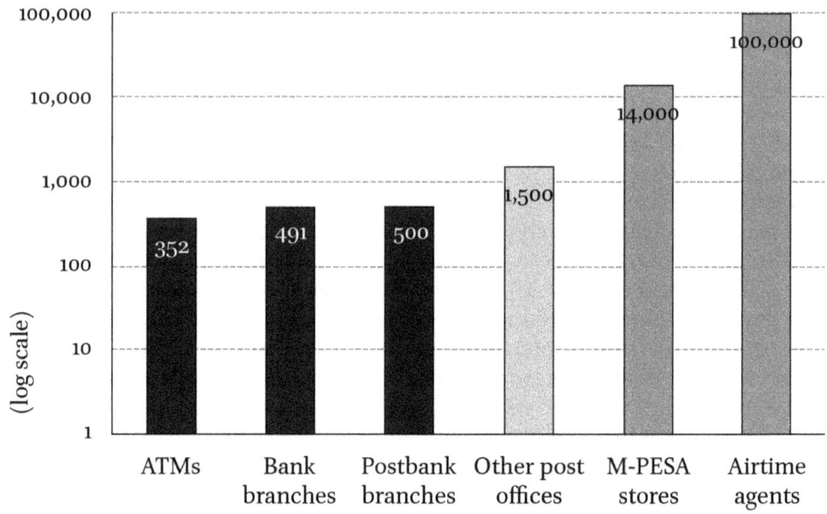

Source: Mas *et al.* (2009)

Figure 5.1 Outlets offering financial services in Kenya

in the lower-income bracket who do not have bank accounts. The Super Agents product has been specifically designed for M-Pesa agents with the aim of increasing e-money availability in the trade and ensuring that agents have an adequate float to service customers effectively. The International Money Transfer (IMT) pilot was set up between Safaricom and Western Union in December 2008 to open up the market for affordable low-value remittances from the UK, with M-Pesa as the cash-out method for registered customers.

M-Pesa is useful as a retail payment platform because it can reach large sectors of the population. Figure 5.1 shows the size of the various retail channels in Kenya. The number of M-Pesa stores is an order of magnitude higher than the total number of branches and ATMs that banks make available to their customers or the number of post offices. Using existing retail stores as M-Pesa cash-in/cash-out outlets reduces deployment costs and provides greater convenience and lower cost of access for users. To establish itself, M-Pesa had to overcome trust and network hurdles as well as obstacles related to its branding, channel management and pricing.

While M-Pesa is gradually encouraging businesses to interconnect, other private-sector players are joining in. Banks, cooperatives and non-bank financial service providers have all connected to M-Pesa to allow existing customers easier access to their accounts. The Equity Bank and Safaricom have partnered M-Kesho services to allow Safaricom customers to deposit money in accounts with the bank. M-Kesho is ultimately forming a stepping stone to formal banking institutions for M-Pesa customers. Phone providers have also launched innovative non-banking services, such as pension funds, mutual funds and insurance, that can only be used via mobile money and that target a low-income market. For example, in addition to keeping their money with Equity Bank, M-Kesho customers are eligible to interest on their deposits and have access to micro credit and short-term loans. The service offers a formal source of both insurance and credit that is extended via the M-Pesa network. This improves individuals' ability to protect themselves against risk by providing better access to emergency loans. The message is that if banks are not in the vanguard, other providers with strong customer relations will step in. Banks, mobile operators and handset manufacturers will all be offering customers mobile wallets. A bank account allows them to earn interest from as little as KSh 1 upwards and other features of the account include micro-credit facilities (emergency credit through M-Pesa), micro-insurance facilities as well as personal accident coverage that translates into full coverage after a year.

In 2008, M-Pesa, in conjunction with the Family Bank, launched the PesaPap financial product so that its 700,000 customers could conduct all their banking transactions without ever visiting a branch. Family Bank believes that the integration between mobile financial services and a personal account is opening up new horizons on the banking landscape. And in October 2008, Safaricom signed a partnership deal with Old Mutual customers to pay Toboa top-ups via M-Pesa. This enables Old Mutual Kenya investors to pay top-ups on their unit-trust investments through M-Pesa (Old Mutual 2008).

Some of the new services provided by M-Pesa have grown due to specific demands from the public. The M-Pesa International Money Transfer (IMT) has become an innovative way of sending money from abroad. As a result of the M-Pesa innovation and its success, Safaricom has received several accolades for its innovativeness, management, positive social contribution, its role in transforming perceptions of Africa in global markets and for displaying high standards of good corporate citizenship. M-Pesa has earned the company national and international recognition from leading networks and organizations in the mobile-phone industry (for example, GSMA) as well as in the non-profit sector (UN Habitat). And it has been given the Kenya Banking Award, the World Business and Development Award, the MSK Best Product Innovation 2008, the Rural Services Award and the Changing Lives Award.

Kajiado County

This chapter considers the adoption and application of M-Pesa services by the Maasai pastoralists of Kajiado County in southern Kenya (Map 5.1). The county (formerly the districts of Kajiado North, Kajiado Central and Loitokitok) is home to some 687,312 people and covers about 21,105 km². The single biggest group, accounting for about half of Kajiado's population, is the Maasai, whose livelihoods have traditionally centred on livestock. This sector is now threatened by land-tenure changes that have triggered social and economic transformations that, in turn, have interfered in the livestock-oriented economy, with water resources being lost to commercial export agriculture. However, other more remote areas of Kajiado are still set aside for livestock keeping and some rainfed agriculture (Rutten 1992, 2008). The other major groups are the Kikuyu and the Kamba (KNBS 2010: vol. 1A).

The overall population density varies but stands at around 20 persons per km² in the central and southern areas, while Kajiado North has higher figures in Ngong Division and fewer people live in Magadi Division bordering Tanzania. Illiteracy here is still among the highest in the country as about 35% of the population (particularly Maasai and women in the southern area) have never attended school, while Kajiado North scores better with a non-attendance rate of only 13%. The level of education here is also significantly higher than in the other two areas where most children did not attend school beyond primary-school level. The Maasai are mainly livestock keepers who occupy the rangelands either as a group of owners or as individual ranch owners. Especially in times of drought, animals are trekked on hoof within the district or beyond as far as Tanzania or Nairobi or they may even be taken by lorry to better grazing locations elsewhere. The nuclear household is mostly young Maasai adults, and employees guard their animals. Another strategy used to overcome the drought is to sell animals to traders that visit the homestead or take them to markets in nearby towns or in Nairobi, Kenya's capital. Besides keeping livestock, Maasai are increasingly engaged in other livelihood options such as small-scale farming and, to some extent, wage employment too.

There are, however, limited opportunities for employment for pastoralists. To earn money, residents sell their livestock at markets and many households depend on remittances sent by family heads and relatives who have gone to the urban centres. There are very few formal financial institutions in Kenya and hardly any villagers have bank accounts. Money transfer services are available in urban centres but residents have

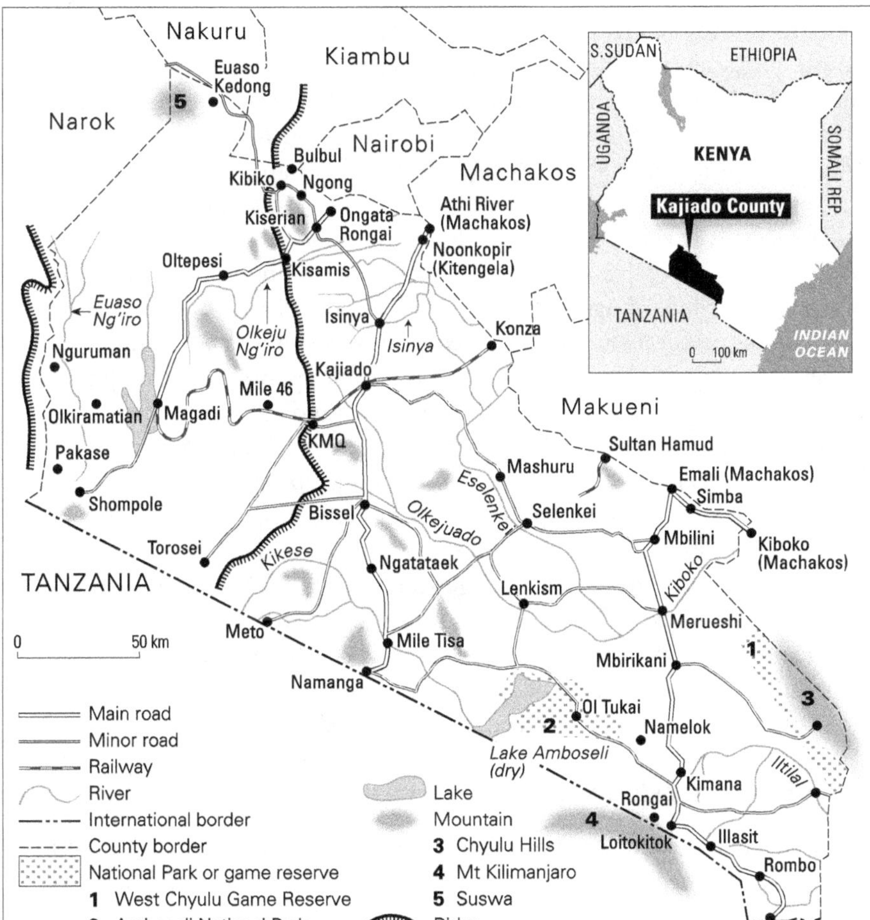

Map 5.1 Kajiado County

to make long trips to retrieve their money from the limited banking facilities, shops or bus companies that serve as conduits.

Interviews were held with 90 Maasai on the effects of the presence of the electronic M-Pesa innovation in the rural areas. The following discussions are based on the findings from field research carried out between December 2011 and February 2012 in three administrative locations in Kajiado County: Elangata Wuas, Isinya and Shompole. A sample of the youth, pastoralist livestock traders and women was taken and they were interviewed using structured questionnaires. The three locations differ in character: Shompole is a remote area with challenging livelihood options, Isinya, by contrast, is about to become part of Nairobi City suburbs and has various services available, and Elanga'ta Wuas falls in between.

The results revealed that, while mobile-phone services are widely used, there is no equivalent use of electronic mobile financial services although there is a huge potential for more absorption of such services. The Maasai pastoralists' habitats, located in some of Kenya's remotest and harshest environments, have a myriad of problems, one of the major ones being financial transactions. As the society is moving from ignorance of a monetary economy to fitting into the rapidly globalizing world, M-Pesa has played a central role in catering for the pastoralists' social and economic wishes. The M-Pesa application is directed largely at the cultivation of livelihood strategies, which help residents to cope with or temporarily adjust to and recover from longer-term shifts in livelihood strategies, stresses and shocks. The facility is also utilized to find and accumulate financial assets and maintain social networks. These services have been evolving and now include the management of weather calamities, which are only too common in this arid and semi-arid environment.

Maasai pastoralists and the M-Pesa technology

Prior to colonialism, the Maasai were among the wealthiest communities in the region, with their wealth being tied up in their animals (Rutten 1992). Disease and drought have decimated their herds and they have suffered the double tragedies of colonial and post-colonial land dispossessions, which have condemned them to rugged and harsh terrains and have disenfranchized them economically and socially. With limited access to land resources, the opportunities and choices of the Maasai have been reduced and today many of them face serious problems sustaining their livelihoods. The information and communication technology (ICT) sector has proved to be an important instrument in improving the performance in livestock marketing and the social life of the Maasai. The mobile phone, for instance, has made a marked difference in this community, probably more so than in many others. Given the distances and poor transport options between homes, the phone has been adopted as part of their lives. For the price of a single goat, mobile phones are becoming an affordable way for these pastoralists to take advantage of modern-day technologies. The Maasai herd owners use them to communicate with herders in the field or to share information on livestock problems, such as disease, losses, water and forage.

Diffusion of the M-Pesa service

The rise in ownership and applications of the phone has provided an impetus for using the M-Pesa service. Of all the respondents interviewed, over 90% of Maasai traders and roughly 60% of pastoralists have mobile phones, which has encouraged them to subscribe to and use M-Pesa. This pattern of technology adoption mirrors that of other products and service innovations, which are often first used by the better-off. The speed at which this service has reached less well-off rural pastoralists and their apparently high valuation of it are unprecedented. Among the 90 households interviewed, 43% were found to have at least one member who had used M-Pesa.

M-Pesa's reach across the socioeconomic spectrum is reflected in a number of dimensions. While the representation of all segments of income distribution in the profile of users has grown, proportional growth has been highest among those at the bottom end of the scale. Households that have taken up M-Pesa recently differ from those who adopted the service earlier. Despite this balancing of the pattern of adoption, M-Pesa remains popular among the well-off, with 93% of households in the top 25% income bracket having at least one user. The same pattern is observed with respect to other indicators of well-being and financial inclusion. Close to 86% of households with bank accounts also have an M-Pesa user, which shows that M-Pesa is not a service targeted at the banked but a service used by them.

The representation of typically poorer, rural households that are less integrated into the broader real and financial economy has also increased. A predictably high share of rural households living in rural towns uses M-Pesa. Similar patterns are observed with regard to educational attainment as better-educated people are more likely to use M-Pesa but growth is, in fact, higher among the less well-educated. M-Pesa is reaching women too. While only 38% of users were female in 2008, the figure had increased to 44% by 2009. Amongst adults over 18 years of age, the share of men using M-Pesa saw a jump from 25% in 2008 to 54% in 2009. But the share of women using the product leapt by comparison from 15% to a level approaching gender parity at 41%. In summary, M-Pesa is one of those rare products that appears to be valued both by those with better opportunities and higher incomes as well as by the less advantaged and the vulnerable. Its adoption would seem to be drawing individual Maasai pastoralists into a financial network in promising ways, i.e. by reducing transaction costs and speeding up business matters.

A major bottleneck in the spread of new technology is thought to be trust in the system. Certainly in the case of finance-related technology, this is of major importance. The M-Pesa system has, to this effect, built in some checks for the sender and the receiver in order to be able to provide proof of the transfers conducted if needed. Each time a sender transfers funds, they receive confirmation with a transaction identification number, the amount sent, the name and number of the person it was sent to, the date and time, and their balance. The receiver gets the same information.

Among Maasai pastoralists, M-Pesa users appear to be becoming more comfortable with the services on offer. The agents interviewed had noticed a steady rise in trust in M-Pesa services. Some shopkeepers claimed that since they became M-Pesa agents they had seen the demand for services surpass that of goods in their shops. They were all in agreement that while only about 65% of the local people reported trusting the M-Pesa agents with whom they transacted in 2008, this figure had increased to 95% by 2009. By the end of 2011, the percentage had risen to 99%. This in turn is reflected in the continued improvements in the behaviour and performance of agents, as is revealed in the findings that show that agents are seen to be performing their jobs according to specified procedures more diligently (95% asked for personal ID in 2009 compared to 70% in 2008) and they are less likely to run out of either cash (5% in 2009) or an e-float (4%) than in 2008 when rates were 16% and 7% respectively. In a rapidly expanding network of agents where the average quality of the pool of potential agents could well be falling, this improvement in agent performance is particularly striking. The advantages of M-Pesa include the fact that there are no monthly service fees, no deposit fees and no costs when paying bills or purchasing prepaid airtime for themselves, their families and friends. And registering and depositing M-Pesa money into an M-Pesa account is free of charge. After making a deposit, a customer is instantly able to transfer money or purchase goods and services. The pastoralists use this money to pay tuition fees, bus fares and make social and economic contributions and payments. Prepaid airtime can be purchased for the account holder or any other subscribers.

M-Pesa was reported to be used to save money by 75% of respondents. Definitions of 'saving' are, however, quite limited and reflect whether an individual or household keeps money on their phone for more than 24 hours. Perhaps the strongest indication that M-Pesa might provide a savings vehicle that facilitates insurance against downside risk is evidenced when interviewees were asked if and why Maasai pastoralists use M-Pesa to save. The share of M-Pesa users who responded that they use it to save for emergencies is on the rise. This new trend might help Maasai overcome periods of stress and assist when opportunities for investment occur. But it is only possible if trust in the system is widespread. Let us now turn to some of the effects that M-Pesa has had on the local Maasai, in particular their livestock traders.

Effects of the M-Pesa service

Maasai pastoralists state that the mobile phone has totally transformed the way they operate their businesses. Many started banking with M-Pesa when the services came closer to them and the benefits were expounded. The M-Pesa system has proved to be simple and intuitive, does not demand much training and one can begin using it immediately after registering. It has become easier for pastoralists to send money, check livestock prices, run other businesses and keep in touch. This reveals just how pervasive mobile banking has become in rural habitats and how the e-money innovation is contributing to the social and economic development of pastoralists in the arid and semi-arid lands.

Only a small percentage appreciate conventional banking services. Residents for years yearned for banking services closer to them and when their immediate financial need was only for maybe KSh 500, they were often forced to withdraw more due to distance and transport costs. When it rains, the journey to the nearest bank could take two days due to unreliable transport and rough roads and spending a night in an urban area usually translates into unbudgeted-for costs. Using M-Pesa to transfer money has had several effects on their household and local economies as well as on their social behaviour. M-Pesa has provided a mobile-wallet revolution for the pastoralists with some surprising results.

Direct monetary gains

The establishing of markets closer to home and the introduction of M-Pesa have considerably reduced the challenges in the livestock trade. Not only do farmers save on travel costs, they also avoid paying middlemen because buyers can now come directly to them. This has increased their profits by nearly 50% per animal. The new marketing development not only benefits the traders but also the women in the community whose incomes have risen after setting up small businesses. As markets continue to grow and improve, more people from the community have started their own businesses, such as animal health services and motorcycle taxis, many of which also benefit from M-Pesa's services. Another unintended outcome is increased peace in the area. Hostilities were more common in the past as men used to fight because they needed each other's products. But with a more integrated economy and easier access to products, the costs of fighting have now become too high.

The M-Pesa service is reputed to be steadily bringing the sense of a money economy to the Maasai pastoralists. For example, in times of drought, livestock buyers walk into homesteads and pay with M-Pesa and then collect any animals they have purchased. For a livestock owner, drought-threatened livestock have more value at this point of purchase than at a market. This practice saves the animals from the long walk to market that can decimate their value and numbers even further. In addition, selling one's livestock before a drought and then restocking when the rains come is becoming a popular strategy among Maasai pastoralists. And community embracement of commercial feeds to save livestock during the most severe droughts is on the rise.

M-Pesa has reduced business running expenses as one can now send cash to wholesalers who then send goods by bus or proxy. There are also groups involved in rural activities, such as charcoal burning, sand harvesting, bee keeping and livestock sales, and M-Pesa is reducing their operational costs and consequently contributing to their growth.

Time advantages

Mobile money makes it easier for people to pay and receive payments for goods and services. People can pay their electricity bills within minutes instead of travelling to often distant offices with a fistful of cash and facing the possibility of having to wait in long queues. Through M-Pesa, consumers can quickly purchase airtime without travelling to the nearest retailer. Other beneficiaries in pastoralist areas include civil servants who in the past would take at least two days to travel to access their salaries.

To achieve this, they had to be absent from official assignments, denying local people their right to be served.

Businessmen too are saving time. For example, suppliers of goods and livestock have devised a system for collecting payments from distributors via the mobile channel and delivering to order points. Businessmen use M-Pesa services to pay off customers who are owed money and when ordering goods from suppliers, one can pay via M-Pesa and have the goods delivered by proxy. Rural markets, shop owners and market traders no longer have to close their shops to travel to pay suppliers and arrange for deliveries. Instead they can send money to suppliers who then dispatch the goods. If a vehicle breaks down in a remote area, pastoralists do not need to track down a bank but can order services or spare parts and pay through M-Pesa. Another practical way that M-Pesa facilitates transport is that it considerably eases payments for public-transport services in the rural areas. Nowadays it is common practice for pastoralists to pay for transport through M-Pesa. Some taxi drivers have also realized that they can operate their businesses more safely with M-Pesa as electronic payments mean that they do not have to carry around large amounts of cash.

Convenience and security

Livestock traders indicated that M-Pesa has become an important part of their livestock business for transferring money and for security reasons. When on a business trip to sell their animals, they deposit all the cash they will require in an M-Pesa account. Those in high-risk areas then detach their SIM cards and only once in town do they withdraw their money from their M-Pesa account, repeating the ritual on the return trip after selling their livestock. Using this facility has meant that pastoralists no longer need to carry cash back from market. Mobile money transfers allow those without a bank account to transfer funds as quickly and easily as sending a text message. In uncertain circumstances, the money on the phone is sent to safe destinations, such as friends or household M-Pesa accounts. Some of those with bank accounts send the money on their phones to banks. Beads and traditional cloth sellers have equally welcomed M-Pesa as they spend an average of KSh 5,000 on transport, food and accommodation when travelling to buy beads, jewellery, ornaments and traditional Maasai *shuka* blankets in Nairobi.

In a country with high levels of violence, robbery, police extortion and corruption, the safety, confidentiality and convenience that M- Pesa offers individuals is welcomed. Pastoralists know that a lost or stolen mobile phone does not mean catastrophe as the thief or 'finder' cannot access their M-Pesa account without knowing their correct personal identification number. Financial protection is thus ensured.

Averting shocks and stresses

The key aspects mentioned above of time saving, banking, security and direct action responses triggered by the availability of M-Pesa have assisted Maasai pastoralists in overcoming individual and/or community stresses and shocks. For example, Maasai pastoralists have innovative ways of using M-Pesa to absorb shocks related to weather vagaries. They can take advantage of the fact that M-Pesa has made it cheaper and easier to send, receive and store money. Pastoralists away from home are able to service drought related shortcomings by sending and making payments for necessities. This has had a profound impact on the ability of pastoralist households to manage their money and improve the way they deal with shocks. Perhaps the strongest indication that M-Pesa might provide a savings vehicle is that it facilitates insurance against downside risk and when asked if and why they use M-Pesa to save, users claimed they use it for emergencies (illness, drought, loss of livestock). There was a rise of 50% among those registered with the service in 2008-2009. Local people feel that the huge losses of livestock in times of stress can be better controlled now they have M-Pesa at their disposal. They have learnt that they can keep their livestock for the three Ms – milk, meat and money – rather than for sentimental reasons. The arrival of M-Pesa has introduced the possibility and reality that one can sell livestock and keep money on one's own phone and restock when appropriate.

It also allows businesses to coordinate their activities better and has opened up a connection to the financial system for a large group of those previously excluded from it. M-Pesa has increased the opportunity for friends and families to make immediate, real-time transactions to help prevent unnecessary periods of hunger and a lack of medical care. Households with M-Pesa are more successful at weathering negative events and do not reduce their food consumption when facing a shock. Additionally, being able to access money quickly allows a family to take a relative to see the doctor on the first day of an illness, and children are able to attend school when they would have otherwise been forced to miss classes or drop out altogether due to a lack of money to pay their school fees.

M-Pesa has helped pastoralists cope with drought situations that threaten their ability to graze and water their cattle. With the help of

Box 5.2 Kashinko ole Ntekese, livestock trader

The advent of the electronic money transfer (M-Pesa) has revolutionized the way I conduct my livestock trade. I used to carry a lot of money in my leather pouch strapped to my body. Later I bought a tailor-made long coat with a deep pocket running to the seams, all in the name of safe money storage. Still, in either precautionary measure, the money was susceptible to theft. Many are the times when I was waylaid on the way from the market or in lodgings and lost the hard earned cash. This always took me back to the begging bowls to restart my business. Livestock trade was all through hard cash. With the coming of the M-Pesa, I do not need to carry money but my phone. Where I have cause to feel insecure, all I need to do is to remove the phone card and hide it somewhere. If I lose the handset, I am assured of the data on the phone card. M-Pesa allows me to buy livestock even from distant sellers. The seller only needs to indicate the value and I pay electronically. I can also contract cattle trekkers who sell my livestock in the market and electronically send me the money while I am in the wilderness. When away from home and my livestock get sick, I am able to buy drugs electronically from the local agrovet shop, or pay a veterinary doctor to treat the animals. I am also able to pay for livestock fodder for the livestock at home in times of drought. I feel very convenienced to access my bank account from anywhere and, at any time of the day for withdraw and deposit with the use of M-Pesa, which provides maximum confidence. As such, the values of the M-Pesa are immeasurable. Overall, my livestock business has been transformed by the advantages of M-Pesa, tripling my income. The safety and speed at which business money moves has been very attractive. This has raised trust in my business.

mobile devices and information shared by various sources, herders can now find out where to graze their cattle, and pay for water and pasturing directly. Pastoralists also use their phones for trade to find the best deals when selling or trading cattle. They use them to engage human and livestock doctors or get medical help, which may not be readily available in remote villages.

In response to the drought in northern Kenya in 2011, a nationwide fundraising effort for famine relief was set up using M-Pesa, with Maasai pastoralists making their own contributions from their remote villages. They know that one day they too might be the beneficiaries of a similar programme. In periods of economic or political instability, pastoralist migrants that have settled in towns look to their rural relatives for support as well. Several of the urban migrants interviewed claimed that they had returned to their villages at times when business was slow and they had run out of cash. It was then difficult to stay in destinations they had migrated to as they could not afford life there, at least temporarily. During the 2007 post-election tensions, urban migrants looked to their villages for support too, receiving both money and mobile-phone credit from their rural relatives.

Boosting the introduction of new devices

Mobile phones and M-Pesa are opening up new ventures in pastoralists' rural environments. Those who live in more remote areas are recharging their phones using solar panels provided by private investors and NGOs, which is creating new business opportunities. Some have bought recharging gadgets to fit the demands of the new technology. The advent of M-Pesa is stimulating local economies by reducing the cost of transactions and giving people access to their own money in their own neighbourhoods. As cash is now available in rural locations and small towns, people are tending to spend more money locally, which is enhancing local economies. In the sustained management of rural water supplies, there have been proposals to apply pre-paid services with the Solar Powered System, which is proving popular because of their low running costs. The Automated Pre-Payment System provides 100% revenue collection and accountability with its automated vending, which is based on mobile money.

Challenges and negative implications of M-Pesa

Traditionally, most Maasai pastoralists have used informal channels to transfer money to family and friends, bus companies and the post office. However, complaints have at times arisen about the channels being both risky and costly. M-Pesa has provided the answer to this. In some instances, respondents claimed not having used a mobile phone in the past but relatives and friends had bought them a phone so they could receive mobile money. A number of women indicated that their sons bought them phones so that they could receive money directly but some of them have not been able to use M-Pesa because they are not well versed with the technology. They also complain that M-Pesa is reducing the number of visits their husbands make to the family home.

Pastoralists' attitudes towards M-Pesa

To Maasai pastoralists, mobile-phone accounts are considered the greatest innovation of present times. They have transformed banking because people no longer require physical bank accounts and an ATM to access

Photo 5.2 Use of M-Pesa at a livestock market. [Photos: Anthony Mututua]

Photo 5.3 A society embracing e-money

money. The connection between mobile money transfer and financial services infrastructure presents the potential for even more services. M-Pesa has shown the diverse ways a mobile phone can be engaged not only for voice and data purposes but also as a banking tool. An accessible and intuitive tool with an unprecedented take-up rate has made M-Pesa a blueprint for other operators.

On the whole, ICT is gaining ground in rural regions in the developing world. This makes Africa, with its 5% GDP growth rate and an abundance of natural resources, a continent on the cusp of immense opportunity and growth. It is now being targeted by multinationals and entrepreneurs looking for investment opportunities. As many African countries bypass telecom land lines in pursuit of a completely mobile infrastructure, sound opportunities abound for addressing the expanding mobile-subscriber base. The rural regions of some African countries would even seem to have more efficient networks than many European countries (Rural21 2011), with visitors surprised to find traditionally dressed Maasai pastoralists in remote villages using mobile phones to catch up on the latest livestock market prices. ICT services have provided pastoralists with a wide range of opportunities from market prices and weather forecasts to financial services.

The Maasai pastoralists' stance on M-Pesa demonstrates its popularity in their community, with 85% of users saying they are 'happy', 'very happy' or 'extremely happy' with the service. The same percentage uses it once a month or less frequently. In other words, they use it mostly for transferring money, something most pastoralists only need to do infrequently. This might suggest that M-Pesa still has some way to go in penetrating other areas of pastoralists' financial lives but it would appear in fact to be happening already among the larger-than-expected cohort.

Far more pastoralists are storing money in their M-Pesa wallet than was initially expected. About 25% say they use it to store money, typically for emergencies or short-term needs, and a further 20% of users report having encountered problems when withdrawing money from an M-Pesa agent, usually because the agent did not have enough cash available. This may be a threat to the M-Pesa system as it is difficult to understand how a bank can survive if one in five of its customers cannot access their money when they need it. But the high levels of satisfaction with M-Pesa suggest that consumers are willing to put up with a certain number of mistakes and limitations because customers have different (lower) expectations of a money transfer service than they do of a bank. Perhaps M-Pesa is such an improvement compared with the alternatives that the net gain in

Box 5.3 Naisenya ene Ngoitoi, housewife

It is interesting that part of my dowry has been paid through M-Pesa. My husband, who works as a policeman and is often away from home, had to agree on the number of cows he needed to pay to have me as his wife. This was mutually converted to cash form and an agreement made with my parents, to whom the money would be paid in instalments. He has been paying the dowry electronically and my family is very happy with this. My husband uses the same method to pay for our children's school fees and the money is sent directly to the schools. The implication of this is that I do not need to travel to urban places to collect money from the post office and take it to the schools, an exercise that was costly and risky. At a personal level, I sell beadworks in the market. To acquire the beads, all I need to do is send money to a supplier in Nairobi who gives the consignment to the next bus travelling to my destination. That bus is also used to deliver my beadwork products to customers who, on receipt of them, send the money to me via M-Pesa. This has really changed my life as I have spare time to take care of my social and economic obligations in the rural areas. I also belong to a women's welfare group where we give money on a merry-go-round basis. If I am not able to physically appear at the gatherings, I send the money electronically to the group treasurer or members whose turn it is to receive the cash. At times my husband is broke and requests some cash from home. All I do is sell a goat, load the money on my phone or have the buyer load the amount on my phone and send it to the husband to salvage whatever the situation may be. The practice gives me some confidence as my husband can transact with me directly without going through intermediaries. In a patriarchal lifestyle like ours, the responsibility that is bestowed on me with the use of M-Pesa for our household management lifts my status at a household and societal level.

usability, security and affordability makes the occasional inconvenience concerning agents worth bearing. Data indicate that customers typically put up with problems. Only one in four indicated having had a problem and attempting to report it to Safaricom. Of the Maasai pastoralist users interviewed, 90% said they did not know how to complain to Safaricom and would ask agents or friends for help.

Conclusion

While the use of mobile phones to transfer money has been a breakthrough for most Kenyans, the system has been a lifeline for rural Maasai pastoralists who are using it in the remote, semi-arid areas of Kajiado County for mobile banking. M-Pesa is changing lives and has the potential to provide low-cost, easily accessible financial services for pastoralists. As the area is served by very few banks, M-Pesa offers a chance for brisk business through mobile banking. The livestock traders and pastoralists themselves at times have immense cash reserves from various sources but, until recently, faced high insecurity and non-existent banking services.

The rapid adoption and frequent use of M-Pesa have engendered a variety of positive outcomes among Maasai pastoralists, as well as unintended consequences. Specific design elements of the M-Pesa system are shaping these impacts. Most importantly, by allowing money to flow electronically rather than physically, M-Pesa lessens and in some cases even eliminates many of the spatial and temporal barriers to money transfers. This increases money flows and allows them to penetrate the remote rural areas where cash is difficult to access. As M-Pesa reaches a critical mass of users, network effects are beginning to develop. Each new M-Pesa user has the potential to tap into an extensive network of potential remitters and lenders. Many rural residents have realized this potential and are drawing on the network to increase their financial inflows. There are also some unexpected consequences of M-Pesa, particularly in savings behaviour to overcome stressful periods such as droughts.

According to pastoralists, the phone has become an ideal mobile wallet that serves as a bank. M-Pesa has increased the ease with which money can be moved around without having to carry cash. The pace of development in the geographical area is not going to see banks willingly locate to the rural areas but it can be construed that the mobile wallet is here to stay. People who were too poor to open a bank account can now have a transaction account at their disposal. The service is being used to make payments for lower-value goods and services, thus increasing the speed of business transactions and lowering costs for users in terms of money, time and risk. An urban pastoralist who used to send money to his nuclear family, other relatives or anyone else using the post office, bus services or friends travelling home can now remit funds quickly and safely for a small fee. And instead of travelling to pay school fees or to cater for unforeseen expenses, parents can make payments to teachers or friends by sending them the money electronically.

Interviews with development agencies show that the rapid uptake and extensive reach of the application has led many to argue that M-Pesa and similar applications have the potential to become more transformational (Rutten *et al.* 2012). Such services provide the foundation for economic development and, in particular, allow the poor to climb onto the banking ladder, thus facilitating access to the formal economy. Furthermore, it is asserted that through such participation, the poor can increase their wealth, diversify their asset base and become more resilient to shocks. Besides reducing vulnerability, the M-Pesa application also facilitates the generation of additional income. Rural households with access to M-Pesa are now better able to protect themselves against downside risks associated with losing their job, livestock or business failure and poor health.

As more Maasai pastoralists become connected by mobile phone, it is not hard to imagine them embracing other applications, such as weather reports. Perhaps a mobile application that provides e-Bay-like services would help Maasai traders to buy and sell cattle more efficiently by finding the best deals in remote villages without them even leaving their homes.

References

CBK (2011) *Annual Report 2011*, Nairobi: Central Bank of Kenya.

CCK (2008) *Annual Report 2008*, Nairobi: Communications Commission of Kenya.

—— (2010) *Quarterly Sector Statistics Report: 2nd Quarter Oct-Dec 2009/2010*, Nairobi: Communications Commission of Kenya.

—— (2012) *Quarterly Sector Statistics Report: 2nd Quarter Oct-Dec 2011/2012*, Nairobi: Communications Commission of Kenya.

Daily Nation (2011) 'M-Pesa Deals Hit 285 per Second', 28 December.

Donner, J. (2007) 'The Use of Mobile Phones by Micro Entrepreneurs in Kigali, Rwanda: Changes to Social and Business Networks', *The Massachusetts Institute of Technology Information Technologies and International Development* 3(2): 3-19.

Duncombe, R.A. (2009) 'Assessing the Potential of Mobile Payments in Africa: Approaches and Evidence from Uganda', Development Informatics Working Paper Series: http://www.cck.go.ke/regulations/sector_legislation.html.

Duncombe, R. & R. Boateng (2009) *Mobile Phones and Financial Services in Developing Countries: A Review of Concepts, Methods, Issues, Evidence and Future Research Directions*, Development Informatics Working Papers 37, 1-33, Institute for Development Policy and Management, University of Manchester.

Kenya National Bureau of Statistics (2010) *Kenya 2009 Population and Housing Census Highlights Brochure: Summary of Census Results*: http://www.knbs.or.ke.
Mas, I. & A. Ng'weno (2009) *Three Keys to M-Pesa Success: Branding, Channel and Pricing*, Nairobi: Bill & Melinda Gates Foundation.
NESC (2007) *Kenya Vision 2030, A Competitive and Prosperous Nation*, The National Economic and Social Council of Kenya: http://www.nesc.go.ke.
Old Mutual (2008) Old Mutual Launch: http://www.safaricom.co.ke/fileadmin/M-PESA/Documents/08.Old_Mutual_Launch.pdf.
Rural21 (2011) 'ICT: A Miracle Cure for Rural Economic Development?', *The International Journal for Rural Development* 1(45): 19-23.
Rutten, M. & M. Mwangi (2012) 'M-Pesa Field Household Surveys among the Maasai Pastoralists of Kajiado County', Unpublished Field Survey Report.
Suri, T. & W. Jack (2010) 'The Economics of M-PESA: An Update': http://www.mit.edu/~tavneet/M-PESA_Update.pdf.
Vodafone (2005) 'Africa: The Impact of Mobile Phones: Moving the Debate Forward', Vodafone Policy Paper Series No. 2: www.vodafone.com/etc/medialib/attachments/.

6

From Gao: Sawaba and the politics of decolonization and insurrection in the Songhay Zone of Mali and Niger (1957-1964)

Klaas van Walraven

The Sawaba Movement headed Niger's first autonomous government during the last years of French suzerainty. Focusing on an end to colonialism, it stood for societal transformation and the uplifting of down-trodden groups. This programme was enhanced by agitating on local issues and defending the cause of specific groups or social categories and integrating these in Sawaba's project. Sawabists also cast their eyes across Niger's frontiers as attempts were made to expand into eastern Mali, making use of identities and communities that straddled the border or had migrated into Niger as a result of colonial development. These social-cultural factors made Mali's Gao region into a natural hinterland, using the innovation of the colonial frontier for its own purposes. This chapter explores these developments from the perspective of the Westphalian notion of statehood and considers how this European innovation, introduced as a result of colonial rule, was adapted in the course of Sawaba's politics in the 1950s. This transformed the border region of eastern Mali and western Niger into a category of its own, i.e. a transitional buffer area where political forces competed for support and control. With the failure of Sawaba's insurrection in the mid-1960s, the frontier 'hardened', with the innovation of Westphalian statehood and its clear spatial frontiers being adapted once again or, rather, being fully implemented for the first time.

Introduction

The establishment of the colonial state in West Africa was, in the long run, to prove to be of great importance to the region's societies. Yet its immediate significance, including that of its post-colonial successor after World War II, is not easily gauged since the key notions of the (post-) colonial state were given new meanings in the course of adapting to their new circumstances. At first sight, it might seem that the colonial state constituted a profound adjustment to the realities of politics as the West African region had experienced them in preceding centuries. After all, colonial territories were established on the basis of the Westphalian notion of statehood, involving spatially defined jurisdiction over all those present on state territory and, concomitantly, the erection of geographical boundaries, visually defined on the ground, to control the comings and goings of people. By contrast, the source of power in Africa's pre-colonial polities had generally been control over people rather than over a geographically delimited territory. As a consequence, the spatial borders that had marked off these polities were not usually made up of sharply defined, let alone recognizable, lines on the ground but were buffer zones where competing authorities exercised varying degrees of control.[1]

This had also been the case in the regions of the Niger River inhabited by people of the Songhay ethnic group, whose settlements stretched from present-day central Mali through what is now western Niger and well into the southern regions of the Nigérien state, up to and including parts of northern Dahomey (Benin). These latter two areas were inhabited by Dendis, a sub-group of Sonrhaï-speaking communities. All these peoples had memories of a joint imperial past as the Songhay polity, centred on the ancient city of Gao, had once been the largest empire in the West African region.

With the imposition of French rule, Songhays found themselves in different colonial territories erected under the pretence (or claim) of Westphalian-style control. The result was that the former imperial capital was now situated in the 'Soudan' territory (the colonial name of Mali) and Songhays were seemingly cut off from each other and, in many cases, from the very heart of a collective, if not uncontested, political past.[2] The new

---

1 This was even the case in such consolidated polities as Asante. The literature on this is too vast to cite here but see, for example, Chazan (1988: 60-97).

2 The Songhay Empire fell after a Moroccan invasion in 1591 and this was followed by centuries of splintered political control. For an introduction to the history of the region, see Dramani-Issifou (1982) and Willis (1976).

political reality was perhaps even more artificial than in other cases of colonial partition since the Songhay peoples of eastern Mali and western Niger[3] were now living in different territories that were ruled by one and the same colonial power. The geographical boundary delimiting the two territories had been imposed simply for purposes of administrative efficiency.[4] Such territorial borders could more easily be changed or done away with, if the colonial power so wished, than boundaries delimiting spheres of jurisdiction between different European powers. And the French did on occasion resort to such measures in other parts of their inter-territorial federation, *Afrique Occidentale Française* (AOF), though this was the exception rather than the rule.[5]

With the onset of colonial rule, new economic developments were introduced that would have a profound effect on the region and affect the reality of the innovation that the Westphalian notion of statehood stood for (or claimed to be) in terms of jurisdiction and border control. With greater physical security, commercial activity increased, gaining new dimensions as a result of technological innovations that, together with the demands of modern administration, brought into being new levels of communication and professions and activities in colonial centres of governance. Transport firms began to move goods and people, clerks and office personnel were trained to fill the lower grades of the administration, and post and telegraph links introduced faster forms of communication. This was also the case in the eastern Mali and western Niger area, especially after 1926 when the capital of the territory of Niger was transferred from the eastern city of Zinder to Niamey, a mere 240 km from the border between the two territories. If small in the beginning, the Niger capital constituted the most important urban concentration in the wider region. (Mali's capital Bamako was located in the distant southwest.) Niamey thus quickly became a destination for numerous people from Niger and eastern Mali.

This process of diffident urbanization led to new social categories, two of which are of relevance here. The so-called *évolués* represented people

3 This chapter focuses on this region and does not discuss the Nigérien-Benin borderland.

4 For a thorough study of Niger's borders, see Lefebvre (2008).

5 The principal changes mainly concerned Upper Volta, which was created as a separate entity in 1920, excised from Côte d'Ivoire, Soudan and Niger. It was then suppressed in 1932 to become part of Côte d'Ivoire and recreated as a territory in its own right after World War II. For the geographical intricacies of the border between Niger and Soudan, see Lefebvre (2008: 228-241).

who had received a Western-style education and achieved middle-ranking positions in the administration or in the European-dominated private sector. The *petit peuple* were rural folk who had also left their villages and now occupied the lowest positions in that same urban milieu: petty traders and artisans, as well as humble employees, alongside technologically more sophisticated workers as drivers, postal clerks and lower-level administrative cadres. If these two categories occupied different positions on the social ladder, both usually belonged to the ranks of the *talakawa*, i.e. commoners, who began to develop a profound conflict of interest with representatives of ascribed authority, namely the chiefs.[6] In fact, several of these people (or rather their parents) had originally been slaves, often in Tuareg communities, but had managed to escape bondage as a result of colonial control that helped shift the balance of power to sedentary communities in the south. Many of these so-called Bellas migrated to the largest city of the region, Niamey, and became known locally as 'Bellas from Gao'. They may not necessarily all have been native to that city but, as their description suggests, had migrated to Niger from the broader Gao area to the northwest. Nor indeed would they all have been of direct slave origin as in practice this name became a generic term for people of low social status, at least among Hausa-speaking people (Baier 1980: 16).[7] Similarly, members of the Kurtey ethnic group, who traced their origins to intermarriage between Peuls and Songhay/Zarma people, migrated from the northwest to more southerly parts (the Tillabéri and Niamey areas), but in their case this happened in the nineteenth century if not earlier. Kurteys settled all along the Niger River Valley, as well as in the capital itself, notably in the Koutoukalé quarter (Bernus ca. 1964: 3). Finally, during French rule, better-educated people from other AOF territories, who collectively came to be known as *AOFiens*, migrated to Niger to fill the more demanding jobs in the administration.

The result was that, by the end of World War II, the Niamey region had a social-cultural composition that was at least in part akin to that of the frontier zone of eastern Mali and western Niger. Even if they had originated from across the border, these people were or steadily became more integrated in the capital of Niger, both in a socioeconomic sense and in terms of their participation in politics. Moreover, the Niger River

6 This was a development that bedeviled social relations and politics in most, if not all, of Africa's colonial territories. For West Africa, see Crowder (1978).

7 In Hausa areas they are known as *Buzu* (pl. *Bugaje*).

continued to dictate a long-distance mode of communication that tended to negate the existence of the land border. In addition, lorries and buses from the Transafricaine company now assured the overland communication and transport of goods and people between Gao and the Niger capital, and with them came the 'travel' of ideas and the expansion of people's political horizons. How did the border between Mali and Niger 'work' in these circumstances? What was its meaning as a Westphalian innovation both in terms of border regime and the concept of jurisdiction, if people not only freely crossed the frontier in pursuit of economic objectives but also took part in the social and political life of the other territory?

This chapter analyses these issues through the prism of a particular political episode and crisis, namely the mode of decolonization that the French imposed on Niger.[8] This was to lead to the marginalization of a trans-frontier social movement that engaged the French in the nature of the devolution introduced, i.e. the establishment of a formally independent Niger that remained, however, under strong metropolitan influence.[9] In the course of this conflict, the members of the movement adapted the Westphalian-grounded border by ignoring it in pursuit of their objectives or by deliberately using it as a protective barrier on the ground in their struggle against the incumbent powers in Niger. Their struggle also shows that the question of jurisdiction, as it is ingrained in the Westphalian innovation, was given a more complex meaning, at least for the duration of the conflict. Cross-border mobilization betrayed forms of political belonging that did not (yet) coincide with the territorial definition of the Niger state, or for that matter that of Mali. The end of this conflict would nevertheless herald the demise of this trans-frontier movement and lead, for a second time, to an adaptation of the Westphalian innovation and its implementation in a fuller sense.

Gao and the politics of decolonization in Niger

In 1954, Djibo Bakary, a charismatic union leader with roots in western Niger, founded a political party, the *Union Démocratique Nigérienne* (UDN) that became known in the country as Sawaba. Its aims were not only to fight colonial rule but also to achieve a broad transformation in society,

[8] For this essentially violent process of intervention by the French, see van Walraven (2009a).

[9] This chapter is based on archival research and fieldwork for my book *The Yearning for Relief: A History of the Sawaba Movement in Niger* (van Walraven 2013a).

agitating, for example, to improve the conditions of the urban proletariat, campaigning against French companies in the interests of petty traders, and challenging the privilege of chiefly authorities and the luxuries of *évolué* life. Finding resonance among different 'classes' of people, including the peasantry, it assumed the hallmarks of a social movement and was the first to bring the politics of decolonization to Niger's countryside. The core of the movement, however, was made up of *petit people*, the rural folk who had migrated to the cities but retained links with the rural areas, and this thus facilitated an expansion of new ideas about justice, society and the good life.

As a result of its core of semi-urbanized *petit peuple*, Sawaba was not a regionally defined party but was spread across the national territory, with fiefs in most urban areas. The party was therefore not only well represented in the central and eastern regions but also in the towns and villages along the Niger River, from the capital all the way to Ayorou on the Malian border. This was partly the result of its union origins and the proximity of Niamey, which enabled activists to spread their message in these areas. The fact that the Bakary family hailed from the area also played a role. Though a Zarma from the Hausa-speaking central region, Djibo Bakary could pose as a 'son of the land' and was thought by many to be a Songhay (these two communities are culturally very close) (Bakary 1992).[10] In addition, many Songhay nobles, who by their social position should have been hostile to a party that thrived on the discontent of social climbers, developed a rapport with Sawaba because they detested Boubou Hama, a Songhay of low caste. An arch enemy of Bakary, Hama was one of the leaders of the rival *Parti Progressiste Nigérien-Rassemblement Démocratique Africain* (PPN-RDA), a party that focused on *évolué* interests and was to rise to prominence with the help of the French.[11]

It was Sawaba, however, that managed to gain political power first when it formed Niger's first autonomous government under French suzerainty in 1957. It owed its strength more to its programme of social transformation and the uplifting of down-trodden groups than to its formal political objectives. The party's political weight was nevertheless enhanced by a willingness to agitate on all manner of local issues ranging from chiefly succession disputes to the plight of (ex-)slaves and to integrate them in its

10 See also van Walraven (2011) on Djibo Bakary.

11 It was also, more broadly, representative of the western region as such, with the exception of towns in the western river valley that formed Sawaba fiefdoms. Niamey was contested territory where the RDA had a majority position.

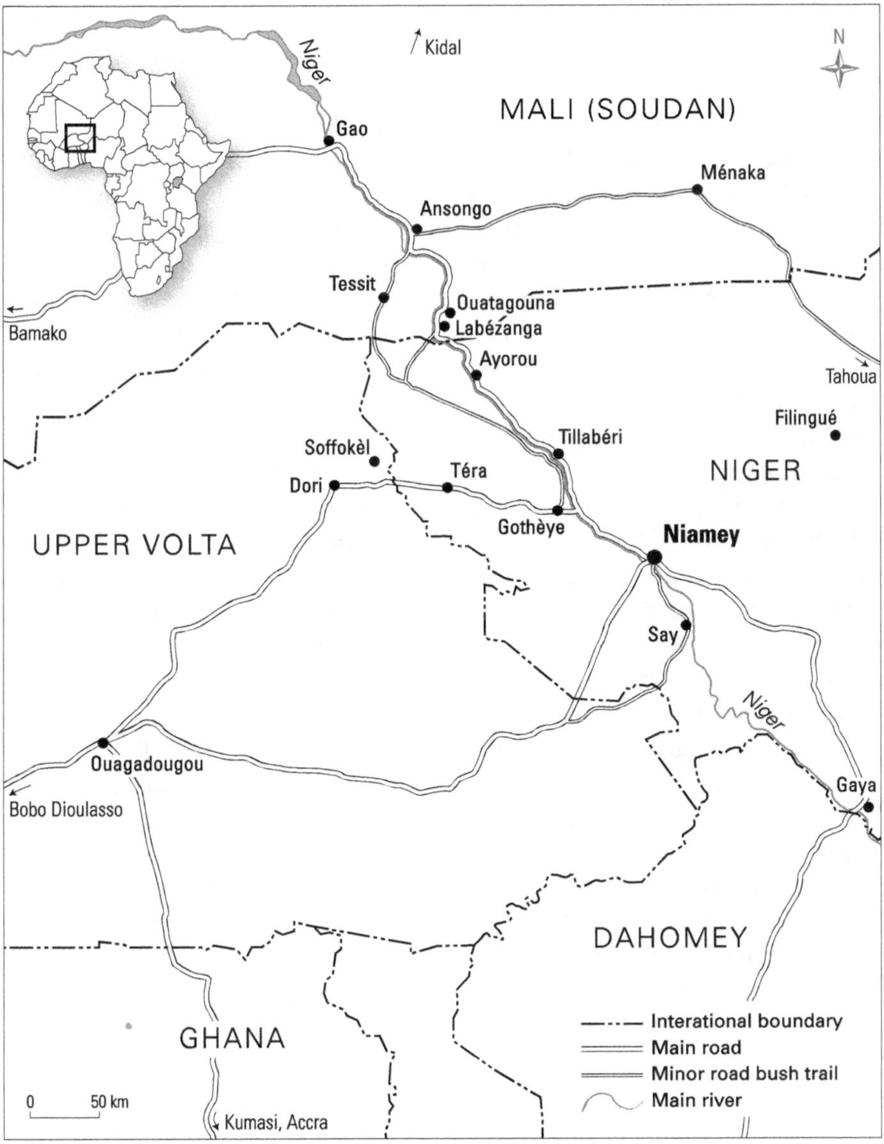

Map 6.1 Gao and West Niger

overall programme. By doing so, Sawaba often ignored Niger's territorial boundaries and, as a corollary, its geographical identity. This was facilitated by the essentially social character of its campaigning, which knew of no spatial borders. As a vehicle for the *petit peuple*, it catered to the interests of people who were in many respects mobile and thus endowed with horizons that went beyond the confines of Niger's territory, and in some cases as far as the frontier zone of Mali and Niger. The people involved included the Bellas from the Gao area but also lorry drivers and workers at Transafricaine who plied the route between Niamey and eastern Soudan. Closer to home in the capital, there were people, often educated *AOFiens*, employed in modern jobs. Though better off than many others, against the background of sharpening political competition *AOFiens* were in danger of becoming marginalized politically as they were increasingly being seen as foreigners in Niger. This xenophobia, which became more prevalent after World War II, arguably points to the gradual growth in territorial identity of the Niger and Soudan polities.[12] In any case, the result was that numerous *AOFiens* were forced to seek a political home, which they found among the Sawabists. Ironically by doing so, they gained access to the highest level of Niger's political arena. As prime minister, Djibo Bakary required skilled manpower, which was in short supply in Niger. He thus appointed several *AOFiens* to his Cabinet, three of whom originally came from Soudan. AOF's existence as a West-Africa-wide colonial federation (though it was to become moribund soon to the benefit of the member territories) still encouraged a broader, inter-territorial focus.[13]

These social, cultural and political facts made Mali, especially its eastern frontier region, a natural hinterland for agitation by Sawaba, which in some ways negated but in others deliberately used the existence of the colonial frontier for its own purposes. It cannot be said that the frontier between Niger and Mali did not exist at all or that attempts at physical control of traffic had no effect but it is obvious that this boundary had not yet 'hardened' in the late 1950s and early 1960s in the sense that clandestine cross-border communication continued and people from across the

12 Though such hatred could also be interpreted against the background of the more common 'stranger' complex in African cultures where later arrivals are not seen as owners of the land or as part of the original population. This points to an awareness of the territoriality of people's lives, though not necessarily in terms of the geographical framework of the post-colonial state, let alone its borders.

13 The federation was dissolved in stages in 1959-1960.

boundary line played roles in agitation at local and even national levels. This underscores that, while sentiments against 'foreigners' were becoming more widespread, the territorial identity of the Niger and Mali states had not yet become the exclusive starting point in the political discourse of the countries concerned. With the inter-territorial federation still in existence, the western Niger and eastern Mali border region represented a category of its own. It was a transitional buffer area where political forces (governments, parties, social movements) competed for support and control.[14] For Sawaba, building a support base in eastern Soudan was potentially useful as it would increase its overall strength for politicking in Niger's political arena but also in case inter-territorial structures survived in one form or another. This was not yet entirely clear in 1957-1958, and Bakary had a passion for federal politics.

In November 1957, a Sawaba delegation led by Hima Dembélé, a unionist from Niamey, travelled to Gao. Accompanied by citizens from Soudan who were resident in Niger's capital, it had accepted an invitation by members of the PSP to visit the city. This *Parti Progressiste Soudanais* was part of an inter-territorial federation of parties that included Sawaba. But it also languished in opposition facing Soudan's ruling party, the *Union Soudanaise-Rassemblement Démocratique Africain* (US-RDA) that, while ideologically closer to Sawaba, was associated via the competing inter-territorial RDA to the PPN, Sawaba's arch rival in Niger. The result was that the arrival in Gao of the Sawaba delegation led to serious trouble. Local (US-)RDA representatives tried to stop the Sawabists from entering the city. When the car carrying Dembélé arrived on the scene, children hurled stones at it and smashed its windows though apparently did not injure its occupants. This was followed by a nightly *tam-tam* organized in Sawaba's honour, which was interrupted by RDA supporters. Other incidents followed in the wake of the delegation's departure. Although the get-together established a Sawaba presence in Gao, Dembélé's delegation was forced to return to Niamey in a battered convoy.[15]

The children in Gao demonstrated that political fiefdoms in the region were conceived of as bastions in the spatial sense and that competitive

14 In recent years scholarly interest in borders and border zones has increased. See, for example, Falola & Usman (2009), Larémont (2005) and Asiwaju (1985).

15 Bordier to France Outre-Mer and Dakar, 2 & 10 November 1957; Centre des archives d'outre-mer, Aix-en-Provence (hereinafter as CAOM), Cart.2198/D.2. A confidential police report is attached to the second letter.

canvassing amounted to trespassing.[16] Yet the altercations also represented a form of agitation that took place at different levels at the same time – local, 'national' and cross-border or inter-territorial – to form a complex web of mobilization patterns that demonstrated that the frontier between eastern Soudan and western Niger did not yet exclude participation in political arenas across the boundary line. One can only surmise as to the nature of Sawaba's reception had the party been affiliated to the inter-territorial framework of the RDA.

This cross-border politicking was seen again the following year but then by way of its mirror image, with people hailing from the Gao region becoming enmeshed in Niger's most violent confrontations to date. In April 1958, PPN-RDA cadres taunted Sawaba's government by trying to establish a presence in the latter's strongholds in the centre and east of the country. This involved propaganda tours that led to street brawls between youths at both ends of the political spectrum, injuries and one fatality that formed the prelude to a settling of accounts in the capital. At the end of the month, Bakary's Cabinet decided to go for a showdown after scuffles had erupted in different parts of the city. This face-off was effected by mobilizing Bellas, most of whom were uneducated and unskilled and employed in menial jobs in Niamey. Archetypal marginals, they belonged to the core of Sawaba's supporters.[17]

> Bella youths were thus loaded onto lorries provided by Pierre Vidal, the French owner of a conglomerate of interests in Niger's building and transport sector, who was Minister of Health in Bakary's Cabinet. Reflecting the involvement of Sawaba's government (government ministers, AOFiens and others, must have been aware of what was about to take place), the Bellas began touring the capital, visiting the compounds of PPN-RDA leaders and engaging party youths mobilized in the PPN's defence. In the ensuing street fighting at least a hundred people were injured. Two died at the compound of Diamballa Yansambou Maïga, a PPN hardliner and Songhay aristocrat who was much hated by Sawabists. A few days earlier, women loyal to Sawaba had tried to attack the Songhay noble, whose roots can reputedly be traced to the rulers of the Songhay Empire. One of the casualties was Maïga's brother-in-law who was killed, possibly in reprisal for shots fired into the Sawaba street mob that killed an adolescent from Gao who had

16 These bastions were not, of course, spatial in a literal sense. It would seem that competitive canvassing would be considered as 'trespassing' if the locality concerned was overwhelmingly loyal to the rival political group.

17 Recueil des principaux renseignements reçus par le Bureau d'Etudes de Dakar pour la période du 20 au 26 août 1959, no. 44 ex. no. 1, 95; CAOM, Cart.3687. Interview with Mounkaila Beidari, Niamey, 2 December 2004.

> recently arrived in Niamey and was employed as a domestic. He was probably Bella. Indeed, Niamey's citizens widely referred to the motorized columns that Sawaba put onto the capital's streets as 'Bellas from Gao', though it appears that they came from both the rural areas close to the capital and Niamey itself. This illustrates that the Gao immigrants, even if they were well established in Niger, were still seen as strangers, i.e. not as Zarma, the area's 'original' inhabitants. Later it was reported that Zarma inhabitants in Niamey killed numerous Bellas in revenge for the attacks and that their bodies were dumped in the Niger River. (Fluchard 1995: 234)[18]

If the Bella-led attacks neutralized the PPN's April offensive, subdued the party and muzzled its parliamentary voice, Sawaba's victory was not to last, though its support in the country grew in subsequent months. As a result of the Algerian crisis and the arrival of General de Gaulle, a radical political transformation was about to unfold in the metropole, leading to the establishment of the Fifth Republic and the constitutional reordering of the Empire. This was put to the vote in September 1958. Sawaba rejected continued autonomy in favour of immediate independence but the metropole was unwilling to cede the strategically vital territory. The French proceeded to tamper with the referendum results and evicted Bakary and Sawaba from power. The PPN-RDA, more accommodating of metropolitan interests, was then put in the driving seat with French support and, when the political dispensation changed again two years later, was allowed to lead the country to a French-controlled independence.[19]

Gao and the insurrection of a social movement

Sawaba's fall from grace in 1958-1959 led to its repression as a legitimate party and union movement. Hated by the French for their agitation and the PPN-RDA for the violent exchanges of the previous years, its cadres were marked men. This was also true for the *AOFiens* who had held senior posts in Bakary's administration. *AOFien* ministers were deported in rapid succession. With the advent of Niger's territorial identity, first under the autonomy regime allowed by the Fifth Republic and then with 'independence', they were increasingly treated as foreigners though this was

---

18 My own account of the Niamey riots is based on a report by a French inspector, Marcel Boyer, who was sent by Paris to conduct an on-the-spot investigation in the second half of May 1958. I/D 2 Juin 1958. Strictement Confidentiel. L'Inspecteur de la France d'outre-mer Marcel Boyer à Monsieur le Ministre de la France d'outre-mer Paris. Objet: Incidents au Niger; I/E 2 Juin 1958; CAOM, FM Cart.1041 & FM Aff. Pol. Cart.2255/D.2. See also van Walraven (2009b).

19 For more details, see van Walraven (2009a).

politically motivated. In March-April 1959, before Niger was even independent, Saloum Traoré and Tiémoko Coulibaly, Sawaba's ministers of social affairs and animal husbandry and both hailing from Soudan, were taken to the border at Ayorou and left to fend for themselves. Diop Issa, Bakary's finance minister, met a similar fate, though he was originally from Senegal. Early in 1960, Maïga Abdoulaye, Sawaba's campaign manager and also from Soudan, was thrown out of Niger with a fellow countryman. This followed other deportations and included people who had often lived (nearly) all their lives in Niger. The extent of the tragedies involved is perhaps best illustrated by the case of Oumar Traoré, an *AOFien* and government clerk of relatively high rank. He was arrested on 20 May 1962 on board a Vidal lorry heading for Gao – completely penniless and only with the clothes on his back.[20]

The *AOFiens* were soon followed by Nigériens themselves – *petit peuple* and allied strata that had supported Sawaba and became the targets of harassment by a regime that stripped them of their jobs, took away their compounds and barred their children from attending school. They had little alternative but to go abroad and so a steady exodus began of those who decided to follow Sawaba's leadership into exile. Bakary himself had been barred from returning after an overseas trip in September 1959 as a result of a politically motivated trial. These exiles underwent military training in Ghana, Algeria and communist China or, with the help of the party, received higher education in the Eastern Bloc, with the communist countries there assisting with scholarships.[21] Sawaba's leadership was planning a comeback by way of a guerrilla campaign and, to this end, needed to build up a military force with educated cadres who could take over the government once the PPN had been toppled.

In this regard Mali also provided support to Sawaba. As already noted, its government was ideologically closer to Sawaba than to the regime in Niger. And with West Africa's territories assuming independence in 1960, the RDA inter-territorial structure finally lost its relevance. Many Sawabists thus made their way along what might be called the 'Mali Trail'. This consisted of at least three different itineraries, two were direct and one had a detour via Gaya on the border with Dahomey. The two direct routes followed the Niger River to Gao or one of several roads leading into Upper Volta from where the journey could be continued to Bamako via

20 B. Moussa, directeur de la Sûreté, to Ministre de l'Intérieur, ca. May 1962; Archives Nationales du Niger (hereinafter as ANN), 86 MI 3 F 12.32.

21 For the relevance of Algerian support, see van Walraven (2005).

Ouagadougou and Bobo-Dioulasso. Many men went on foot, while others took the Transafricaine bus or hitched a ride with one of the lorry drivers plying the road along the left bank of the river. Canoes were used too. Niger's Sûreté reported in August 1961 that cadres were ferried across the river from a fishing village near Niamey, the home of an uncle of Hima Dembélé, who himself played a key role in the evacuation of cadres.[22] The town of Gothèye also played a part here, being located at the ferry crossing between the capital and the border. The fact that Gothèye and other Songhay towns had been pro-Sawaba, which was reinforced by hostility to PPN icon Boubou Hama, made the route along the river attractive. Although the main road to Gao was along the left bank, cadres occasionally crossed to the other bank to follow smaller but safer tracks there. Further upstream there was a possibility to cross the river at Ayorou, where there had been an important Sawaba following in the past. It seems cadres preferred to go to the right bank here to avoid crossing the border by the main road where Niger's regime had established an administrative post to prevent Sawaba infiltration from Gao. It was not generally very difficult to cross the border although it was not without risk. Sawaba's long-standing presence in the river valley, if not in the Gao area itself, in addition to the welcoming attitude of the Malian authorities, made the region a politically integrated territory rather than a forbidding border zone where furtive crossings were impossible.

Sawabists established themselves in several Malian cities: in Bamako, of course, where the authorities provided Bakary with lodgings, but also in a host of other towns. Gao and other satellite towns were however all the more important since they were located close to the border and could assist in the reception of cadres and with infiltration missions into western Niger where the populace was to be sensitized to Sawaba's cause and intelligence could be collected. In October 1959, Djibo Bakary visited the former Songhay capital where he was received by Soudan's Minister of Justice and welcomed by enthusiastic crowds. He held a meeting with government officials as well as several of his own lieutenants, some of whom still lived in Niger but had travelled to Gao for the occasion. Others had already fled Niger or been kicked out, such as Maïga Abdoulaye who now lived in Ansongo to the southeast of Gao and closer to the border, where he coordinated the execution of Bakary's orders from Bamako. The Malian authorities assisted, and this helped consolidate what had initially been a precarious political presence. Malian assistance could even go one

22 Note d'information, 11 August 1961; ANN, 86 MI 3 F 4.4.

fundamental step further, i.e. *across* the formal Niger-Mali boundary line. In the autumn of 1959, an official from the district of Ansongo crossed the border to Ayorou to assure Sawabists of support from the Soudan government. The regime in Niamey was outraged by the visit.[23] It showed that post-colonial identities did not yet constitute the spatial framework determining all political mobilization. The boundary line could still be temporarily ignored if thought expedient, and this demonstrated that the Westphalian innovation, based as it was on a claim to exclusiveness of jurisdiction, had been adapted. In the process, different actors – the Soudan government, the Niamey regime, the Sawaba rebels – enjoyed varying degrees of support in one and the same area.

For Sawaba, Gao and its environs served multiple purposes and challenged the region's post-colonial territoriality. First, the city itself became one of the movement's bases. An office was established, the party archives were to be brought from Niamey and, early on, cadres travelled to Gao to set up an intelligence and propaganda service. Bakary himself established a new political vehicle in Gao, whose statutes he then ordered Sawaba's domestic leader Georges Condat to deposit with the authorities in Niamey, although this failed to materialize. The message of the new party structure was to be spread with the help of trans-frontier traders and transporters, and, to this aim, an important meeting was scheduled to take place in Ayorou just inside Niger. It was rumoured that Bakary would attend in person and while this was improbable, it showed that with Soudan soon to become independent (as Mali), the Gao region would become a strategic hub in the confrontation with the regime in Niamey. In November 1959, the French expressed fears that the city of Gao was 'susceptible of becoming the command post of Sawaba and the place from where all subversive actions towards Niger (would) run'.[24] And towards the end of the year, Sawabists from Gao travelled to Kumasi, in Ghana, where they helped establish a local office. This would assist in the processing of recruits who fled western Niger and were en route to Accra where they were to undergo training or be transferred to other countries.[25]

23 Recueil des principaux renseignements reçus pour la période du 6 au 12 août 1959 & 10 au 16 sept. 1959, no. 42/47; 1 au 7 janv. 1960, no. 62; CAOM, Cart.2249, 2251, 2252.

24 Recueil des principaux renseignements reçus pour la période du 5 au 11 Novembre 1959 no. 55; CAOM, Cart.2251 ('susceptible de devenir le P.C. [poste de commandement] du Sawaba et le lieu d'où seront dirigées toutes les actions subversives en direction du Niger').

25 Recueil des principaux renseignements reçus pour la période du 22 au 28 Janvier 1960, no. 65; CAOM, Cart.2252. Kumasi was also close to military training camps that Sawaba could use in Ghana.

Gao itself was an important centre for the transfer of Sawabists who fled Niger. One of Sawaba's representatives there, a former Niger treasury clerk called Alazi Soumaila, received and vetted incoming refugees before sending them on, usually to Bamako, from where they then travelled to destinations abroad. Other people active in this vein were Maïga Abdoulaye, based in nearby Ansongo, Saloum Traoré, his Cabinet colleague who after his expulsion established himself in Gao and instructed cadres infiltrating Niger's western region,[26] and Farka Maiga, a Sawaba unionist who had worked at Entreprises Vidal in Niamey but was an *AOFien* and probably from Soudan. He was deported with other *AOFien* union members at Vidal's for having instigated a strike to protest Bakary's trial (according to the government). Their expulsion had led to unrest among Soudanais living in the frontier zone, who shared ethnic-cultural bonds with the menial workers in Niger's capital, be they Songhay, Kurtey or Bella.[27]

There was also accommodation in Gao to put up Sawaba's recruits, many of whom then travelled on to Bamako by river ferry.[28] The principal vetting took place in the Malian capital by Daouda Ardaly, a militant who had developed extensive contacts in Eastern Europe. In the former Songhay capital though, Sawaba's presence grew in significance, assisted by the readiness of the Malian government to turn a blind eye if not render the occasional active service to its political allies from Niger. Members of Bakary's family settled in Gao, as did ordinary cadres. Its importance was shown when, in 1961, Sawaba's leadership contemplated the appointment of Hima Dembélé as Gao's *chef de poste*, a plan that was only abandoned when Dembélé was needed in Niamey to assist in the infiltration of Niger's military.[29]

If Mali's eastern region was not Sawaba's principal forward base (it was Northern Nigeria),[30] the Sawabist presence there gradually developed. In

---

26 Ministre de l'Intérieur to Chef des Services de Police, Niamey, 31 May 1961; ANN, 86 MI 3 F 3.8; Recueil des principaux renseignements reçus par le Bureau d'Etudes de Dakar pour la période du 14 au 20 mai 1959, no. 30; CAOM, Cart.3686. Traoré probably died, however, in late 1959/early 1960. *l'Essor*, 15 Jan. 1960; J. Colombani, to Ministre de l'Intérieur, 1 Febr. 1960; ANN, 86 MI 3 F 3.8.

27 Recueil des principaux renseignements reçus par le Bureau d'Etudes de Dakar pour la période du 3 au 9 sept. 1959, no. 46, 20 au 26 août 1959, no. 44; Synthèse politique, no. 1214 CP, Sept. 1959; CAOM, Cart.3684 & 3687.

28 Interview Ali Amadou, Niamey, 28 Jan. 2003.

29 Airgram, Department of State, A-259, 28 March 1963; United States National Archives and Records Administration, II RG 59, POL 30, Niger, 2/1/63, Box 3997; Note d'information, 20 June 1961; ANN, 86 MI 3 F 3.10.

30 In essence this was due to the fact that Sawaba's most important strongholds were in Niger's central region, notably in the cities of Zinder, Tessaoua and Maradi, which were

addition to Ansongo, a satellite base was formed in Tessit, a village 75 km south of Ansongo and close to the Niger border on the right bank of the river, and there was one at Ménaka, much further east that was connected by road to Ansongo and was useful for missions to Niger's Filingué area and further afield to Tahoua in central Niger. The cell in Tessit, where the canton chief possibly supported Sawaba, was initially run by Idrissa Arfou. He was an old party hand who had earlier been in Accra where the party had established a presence as well. He became active in Tessit and Gao as a liaison officer. Tessit, also a stone's throw from the Voltan border, was not only useful for infiltrations of the right embankment of the Niger River but also played a role in connecting Gao with Sawaba's presence in Accra, which was a ten-day journey. By 1963 Arfou was so-called *chef militaire* of Sawaba's local guerrilla unit, while Mounkaila Beidari was Tessit's 'political commissar'[31] and responsible for infiltration and intelligence in the Téra sector, which was part of Niger's western region close to Upper Volta and Mali. By the end of 1963, Beidari was succeeded by Mounkaila Albagna, who undertook several intelligence missions in this sector. Ansongo was handled by Maïga Abdoulaye and although by the spring of 1961 he may have been working for the party from Gao, Ansongo continued to function as a satellite base for the guerrillas. The Ménaka area also had a Sawabist presence and there are indications of propaganda activity there and, possibly, of an arms cache.[32]

Sawaba's leadership divided Niger's western region into three military sectors: Téra on the right embankment, Ayorou-Tillabéri on the left, and further east, Tahoua, which could be reached by way of Ménaka. The right embankment could be infiltrated by heading down from Ansongo and the village of Ouatagouna and crossing the river near Labézanga on the border with Mali or at Ayorou and then continuing along the road to Téra and Gothèye. Alternatively, party workers travelled by canoe downstream.

all close to the northern Nigerian border. In contrast to the situation in Mali, Sawaba's support in northern Nigeria came from the local opposition movement, the Northern Elements Progressive Union (NEPU). Northern Nigeria's regional government was implacably opposed to NEPU and, by extension, Sawaba. See van Walraven (2013b).

31 The second-in-command who also monitored the loyalty of the unit's guerrillas and had to supervise the political organization of Niger's infiltrated regions.

32 *Fraternité-Hebdo*, 16 July 1965; Recueil des principaux renseignements reçus par le Bureau d'Etudes de Dakar pour la période du 24 au 30 sept. 1959; CAOM, Cart.2250; *Ibid.*, 1 au 7 janv. 1960, no. 62; Ministre de l'Intérieur to Chef des Services de Police, Niamey, 31 May 1961; ANN, 86 MI 3 F 3.8; interviews Mounkaila Beidari, Niamey, 28 November 2003 & 15 December 2009 & Mounkaila Albagna, Niamey, 29 November 2003.

Early in 1960 the French reported that 'rather frequent liaisons [were] carried out by canoe along the river'.[33] Most of these missions headed for Gothèye which, benefiting from its Sawaba community and contacts with Ghana through labour migration (a good cover for infiltration missions) was the principal cell on the right embankment. Ayorou played a similar role on the left-hand side of the river, probably maintaining contact with cells such as that at Tillabéri, whose cadres were in touch with Hima Dembélé in Niamey.[34] On the right embankment, missions may also have targeted Téra. Parties heading for Téra used Voltan territory to assemble, among others, at the town of Soffokèl, which is east of Dori and a stone's throw from the Voltan-Niger border, before penetrating Niger.[35]

Since Sawaba's cadres and guerrillas (who became known as commandos) did not amount to a massive presence in the region, much could go on unnoticed or was not considered politically threatening by the Malian authorities. This meant that, while the locals of Tessit, Ansongo and Gao must have been aware of the Nigériens in their midst, their presence did not immediately create problems for Mali's government with the regime in Niamey. Nevertheless, Sawaba infiltration missions in Niger's western region started early on and were aimed at intelligence gathering, sensitizing the populace, contacting cadres of the party's internal wing and constructing arms caches, although these were safer across the border in Mali. In June 1961 a Malian was arrested on the frontier with a list of no less than 150 domestic Sawabists. The French also suspected that arms were being transported by way of the river and a year later, in October, the Malian *gendarmerie* at Labézanga seized a dozen rifles that a Niger national was trying to smuggle across the border although this may have been an example of ordinary arms trafficking.[36]

All this could only have developed because of one essential element, namely the adaptation of the Westphalian notion of statehood with its

---

33 Recueil des principaux renseignements…1 au 7 janv. 1960, no. 62 ('liaisons assez fréquentes s'effectuent par pirogue sur le fleuve').

34 The Tillabéri cell included a Transafricaine bus driver from Gao. Note d'information, 20 July 1961; ANN, 86 MI 3 F 3.10.

35 Interview Mounkaila Beidari, Niamey, 28 November 2003; Daouda Hamadou, Niamey, 20 December 2009; *Le Niger*, 21 June 1965 & 2 November 1964 (containing the text of the interrogation of Robert Seguinikin).

36 Note d'information: Activités de la Sawaba au Niger (Le Premier Ministre. Etat-Major Général de la Défense Nationale. Division Renseignement, 29 June 1961, no. 5.211; Service Historique de l'Armée de Terre, Vincennes (hereinafter as SHAT), 10 T 717/D.2; Bulletin de Renseignements Hebdomadaire, 15-21 Oct. 1962, 25.747/2/SC, no. 25; SHAT, 5 H 121.

exclusiveness of territorial jurisdiction and the state territory as the sole basis of political mobilization. At the time, such an adapted notion could be seen at work in other countries as well, as it constituted an element of 'foreign' policy making in the wake of independence, with new regimes testing the extent of their spatial influence. This was particularly the case with governments driven by left-leaning, revolutionary aspirations, if not expansive ambitions, such as Egypt, Algeria, Ghana and Mali (van Walraven 1999: Chapters 3 & 4). It can thus be observed that, after the accession to independence, frontiers often did not immediately 'harden' to discourage foreign, or rather cross-frontier, intervention. This was the case, for example, along the border between Algeria and Niger and between Niger and Nigeria.[37] In the Malian case, the negation of the Westphalian notion was made possible partly by active commitment as Sawabists en route to the Eastern Bloc were issued with Malian passports, turning the Westphalian idea of jurisdiction on its head, and partly by neglect: with Mali's government based in the southwestern (Bambara) region, Sawaba's political mobilization in the Songhay zone could continue unhindered, building on cross-border solidarities that paid little heed to the formalities of the Westphalian notion. This allowed other actors, such as rebel forces, to build on a trans-frontier socio-cultural reality. At the same time, it cannot be said that the post-colonial frontier did not exist on the ground, since it was through its existence that the Gao region was able to grow into a Sawaba base. In its absence, this would have been much more difficult, as a protective barrier against the Nigérien regime would then have been absent and, concomitantly, Sawaba's activities would have been pushed further underground.

The hardening of a frontier: The decline of Gao as base for cross-border subversion

This situation was able to continue for as long as it did not create serious problems for the Malian government.[38] That the Niamey regime likewise

37 In the latter case, cross-boundary communication, or perhaps 'interference', was carried by social, economic and cultural forces that existed on the ground, rather than by political actors representing regimes, whether they be Nigeria's federal government or the regional government of Northern Nigeria. See van Walraven (2013b).

38 The Malian government did not really benefit from the situation as it reinforced tensions with and mistrust by the Nigérien authorities. The expected benefit would only happen if Sawaba were able to topple the regime in Niamey, thus creating a government there that was potentially grateful to its Malian friends.

had few qualms about violating Mali's territorial sovereignty was shown in May 1962 when a hit man from the regime killed Daouda Ardaly in Bamako.[39] The murder of Ardaly, by then Sawaba's principal recruitment officer, encouraged a development that had already been set in motion, namely the gradual move of Sawaba's headquarters to Accra where the Ghanaian government offered better security. Mali's regime had already tried to improve ties with Niamey, as economic problems pushed Bamako to improve relations with more prosperous countries like Côte d'Ivoire. (Niger's regime was the latter's protégé.) Niger used this to press for greater security in the border zone and in February 1962 the Minister of the Interior met his Malian counterpart in Gao, followed by a meeting at the border of the Commandant de Cercle of Tahoua in central Niger and his colleagues in Ménaka and Kidal to the northwest in July. One of the issues discussed was border patrols. French intelligence noted that the Malians toured the frontier region once or twice a month, while the Tahoua authorities were unable to do so because of a lack of resources. This proved no stumbling block to infiltration but the talks in Gao between Malian and Nigérien leaders on a rapprochement between Niamey and Sawaba (and Bakary's return to Niger) ended in stalemate.[40] Thus, in September of that same year the French noted about the regime in Niamey that it was 'first of all Mali, with its armament, its enlisted youths and the formidable exile it lodge(d) that one fear(ed) the most'.[41]

Preparations for an attack on the regime in Niamey continued unabated, also in the Songhay zone. The whole plan, however, went up in smoke in the autumn of 1964 when the arrest of Sawaba *guerrilleros* in central Niger threatened to destroy the movement's infrastructure and the leadership felt forced to go on the offensive, even though militarily the situation was not ripe. This is not the place to analyse what happened[42] but suffice it to say guerrillas found that the local populace, who had recently been cowed into submission by regime militias, failed to rally to their cause

39 For a discussion of Niamey's responsibility, see van Walraven (2013a: Chapter 11).

40 Premier Ministre. SDECE. Destinataire no. 501, 2 Nov. 1961; SHAT, 10 T 717/D.2; Recueil des principaux renseignements collectés au cours de la période du 12-18 avril 1962, no. 1; SHAT, 5 H 91.

41 Bulletin de Renseignements Hebdomadaire, 3-9 September 1962, 25.547, no. 19; SHAT, 5 H 121. In due course, however, it was from Ghana that the principal threat came, as most of Sawaba's guerrillas came from there when the 1964 offensive (see below) was launched (*'tout d'abord le Mali, avec son armement, ses jeunesses embrigadées et le redoutable exilé qu'il héberge que l'on craint le plus'*).

42 See van Walraven (2013a: Chapters 12 & 13).

in practically all regions. This was particularly true in the central Hausa-speaking region: since the regime suspected Sawaba's support there to be the strongest, it had made a special effort to intimidate its local population.

It was therefore in the western region that support for Bakary's men manifested itself most clearly in the course of the guerrillas' arrival.[43] This corresponded with the substantial support base built up in the area in the past, as well as the hostility of the local Songhay chiefs *vis-à-vis* government notables in Niamey. Consequently, the French reported that the population remained passive regarding regime mobilization and that some guerrilla units were hard to catch. The problem, however, was wider than the Songhay community alone, since the French (in their ethnicized perspective of Nigérien politics) were surprised to note that the population in the Téra region 'although of the Zarma race, only reluctantly gave their support to the forces of order'. French intelligence linked this to hostility towards the person of Boubou Hama, though Diamballa Maïga was hated here too.[44] When the Sawaba invasion backfired, it was in the areas of western Niger that the regime's repression proved most brutal. PPN militias were dispatched to the Say and Téra regions 'to give back to those populations "a more developed civic sense"'. French intelligence clarified this by adding that 'numerous arrests [were] carried out in these regions'.[45] In towns like Dargol, Bandio, Gothèye and Téra, which were widely seen as Sawaba fiefs, inhabitants were forced to denounce each other and were beaten up and maltreated. They also witnessed several executions that were either extra-judicial or took place with the authority of the *Cour de Sûreté*.

The horrors in the western region drove home the fact that Sawaba was, contrary to what some have claimed (Idrissa 2001: 64), more than a mere 'Hausa' movement. Its strength here was in part bound up with cross-border (cultural) solidarities. These provided Niger's west and Mali's Gao region with a degree of integration that gave a meaning to the territorial border that was different from the Westphalian notion ostensibly

43 Also in a few other areas such as the Lake Chad region and Tahoua.

44 Untitled document, n.d., SHAT, 10 T 717/D.2; Evénements survenus en Afrique Francophone pendant la semaine du 2 au 8 nov. 1964; SHAT, 10 T 210; Premier Ministre. SDECE. Destinataire no. 541, 24 Nov. & 8 Dec. 1964; SHAT, 10 T 719/D.2; interview with Mamoudou Béchir, Dargol, 31 Oct. 2005 (*'pourtant de race djerma, ne prêtent qu'avec réticence leur concours aux forces de l'ordre'; 'contrairement à ce qui s'était passée sur les frontières Sud'*).

45 Premier Ministre. SDECE. Destinataire no. 541, 24 November 1964.

underlying it. It existed as an obstacle to communication but also acted as a channel for all sorts of cross-border contact. Its existence, and that of the two different post-colonial states that it defined, proved at the same time the *sine qua non* for the development of Sawaba's power base in Gao.

The consequences of Sawaba's failed invasion were, however, to change all this. French intelligence noted that the executions in Niger's western region were 'hardly appreciated' by the population,[46] but in the long run this was of little help. The failed invasion radically altered relations between the governments of Mali and Niger at the national level, as Bamako felt constrained to take notice of what was transpiring in the distant east. Famine conditions in Mali and the need for food imports, from Niger among other countries, provided the leverage that Niamey needed to force Mali to respect the border and, with it, the notions of sovereignty and jurisdiction in ways that conformed more to the Westphalian ideal. Sawaba guerrillas were arrested and handed over at the frontier. After their studies in the Eastern Bloc, Sawaba students saw their Malian passports confiscated, which forced them to surrender their fate to Niger's *Sûreté*, a mirror image of the fate suffered by Niger's *AOFiens*. In the process, the border between the two West African states was transformed or 'hardened' into something conforming more closely to the original Westphalian innovation that was meant to underlie it. In a way, the Westphalian notion was now being fully implemented for the first time. This helped consolidate post-colonial regimes that, on both sides of the border but certainly to its east, were marked by a lack of political freedom. Ironically, this too formed part of the new political tradition, with statehood in both countries now striving for an African variation of the *cuius regio, eius religio* that had originally inspired the Westphalian 'peace'.

Conclusions

The politics of eastern Mali and western Niger became closely intertwined in the post-World War II era, based on a myriad of cultural, social and developmental factors such as migration to the city and the evolution of modern transport, which helped to underpin common political positions. This also represented a novel way of employing and, in the process, transforming the Westphalian notion underlying the colonial state and its territorial borders, using the border as a channel for communication and, at

[46] Premier Ministre. SDECE. Destinataire no. 541, 8 December 1964 (*'guère appréciée'*).

the same time, as a protective barrier that allowed the build-up of a power base for cross-border incursions.

In the process, the novelty of Westphalian-style jurisdiction, based as this originally was on a territorially dictated exclusiveness, was transformed in such a way that it allowed rival forms of belonging and mobilization. However, while these forms of belonging and mobilization allowed simultaneous engagement in Malian and Nigérien politics, the territorial border between the two states was never called into question. The eastern Mali and western Niger region may have represented a transitional buffer area where rival political forces competed for support but this did not amount to a buffer zone as in pre-colonial times, since formal claims to jurisdiction were not disputed. The handing out of passports had been a matter of expediency, not principle.

The failure of the Sawaba rebellion was to herald in a new era where the border was again transformed, this time into an entity assisting political interaction at national and not cross-border level. The result was that the innovation of Westphalian statehood was transformed once again but now, paradoxically, to conform more to its original formulation.

References

Asiwaju A.I., ed. (1985) Partitioned Africans: Ethnic Relations across Africa's International Boundaries 1884-1984, London/Lagos: Hurst/University of Lagos Press.

Baier, S. (1980) *An Economic History of Central Niger*, Oxford: Clarendon Press.

Bakary, D. (1992) *Silence! On décolonise: Itinéraire politique et syndical d'un militant africain*, Paris: L'Harmattan.

Bernus, S. (ca. 1964) *Niamey: Population et habitat; documents d'enquêtes*, Etudes nigériennes 'Documents' 11, IFAN/CNRS.

Chazan, N. (1988) 'The Early State in Africa: The Asante Case'. In: S.N. Eisenstadt, M. Abithol & N. Chazan, eds, *The Early State in African Perspective: Culture, Power and Division of Labor*, Leiden: Brill, pp. 60-97.

Crowder, M. (1978) *Colonial West Africa: Collected Essays*, London: Cass.

Dramani-Issifou, Z. (1982) *L'Afrique noire dans les relations internationales au XVIème siècle: Analyse de la crise entre le Maroc et le Sonrhaï*, Paris: Karthala.

Falola, T. & A.A. Usman, eds (2009) *Movements, Borders and Identities in Africa*, Rochester, NY: University of Rochester Press.

Fluchard, C. (1995) *Le PPN-RDA et la décolonisation du Niger 1946-1960*, Paris: L'Harmattan.

Idrissa, K. (2001) 'La dynamique de la gouvernance: Administration, politique et ethnicité au Niger'. In: K. Idrissa, ed., *Le Niger: Etat et démocratie*, Paris: L'Harmattan, pp. 15-83.

Larémont, R.R., ed. (2005) *Borders, Nationalism, and the African State*, Boulder, CO: Lynne Rienner.

Lefebvre, C. (2008) 'Territoires et frontières: Du Soudan central à la république du Niger 1800-1964', PhD Thesis, University Paris I.

van Walraven, K. (1999) *Dreams of Power: The Role of the Organization of African Unity in the Politics of Africa 1963-1993*, Aldershot: Ashgate.

—— (2005) 'From Tamanrasset: The Struggle of Sawaba and the Algerian Connection (1957-1966)', *Journal of North African Studies* 10(3-4): 507-527.

—— (2009a) 'Decolonization by Referendum: The Anomaly of Niger and the Fall of Sawaba, 1958-1959', *Journal of African History* 50: 269-292.

—— (2009b) 'Vehicle of Sedition: The Role of Transport Workers in Sawaba's Rebellion in Niger, 1954-1966'. In: J.B. Gewald, S. Luning & K. van Walraven, eds, *The Speed of Change: Motor Vehicles and People in Africa, 1890-2000*, Boston/Leiden: Brill, pp. 75-103.

—— (2011) 'Djibo Bakary'. In: H.L. Gates & E. Akyeampong, eds, *Dictionary of African Biography*, Oxford: Oxford University Press, Vol. 1, 355-357.

—— (2013a) *The Yearning for Relief: A History of the Sawaba Movement in Niger*, Leiden/Boston: Brill.

—— (2013b) 'From Kano: Clandestine Politics in Niger and Nigeria and Sawaba's Cross-Border Ties with NEPU (1958-1966)'. In: A.S. Awenengo & C. Lefebvre, eds, *Boundaries and Independences 1940-1965*, Basingstoke: Palgrave Macmillan.

Willis, J.R. (1976) 'The Western Sudan from the Moroccan Invasion (1591) to the Death of Al-Mukhtar Al-Kunti (1811)'. In: J.F.A. Ajayi & M. Crowder, *History of West Africa*, Vol. 1, London: Longman, pp. 512-554.

7

From self-help group to water company: The Wandiege Community Water Supply Project (Kisumu, Kenya)[1]

Samuel O. Owuor & Dick Foeken

Water services in urban Africa are generally in a bad state and low-income neighbourhoods are not usually connected to the municipal water supply and infrastructure. The Kenyan government attempted to address these problems with its Water Act of 2002 and by creating water and sewerage companies. This chapter recounts how this idea was adopted and transformed by people in the Wandiege neighbourhood of Kisumu town who improved their local water and sanitation situation by setting up their own water supply system in an innovative way.

Introduction

Water is a basic human right as it is fundamental to life and is vital for socio-economic growth and development at all levels, from the national level to that of the individual. Access to water (and sanitation)[2] is a key factor in improving people's health, economic productivity and social well-being as both social and economic activities rely heavily on the quantity and quality of the water available. Access to water is thus an

[1] Comments by Moses Mwangi, André Leliveld and Iva Peša on an earlier version of this chapter are highly appreciated.

[2] This chapter predominantly focuses on water but sanitation issues cannot be totally ignored.

essential component of any attempt to alleviate poverty. Yet in achieving this target, 'Sub-Saharan Africa remains the area of greatest concern. It is a region of the world where, over the period 1990-2004, the number of people without access to drinking water increased by 23%' (WHO/Unicef 2006: 3). Moreover, the region experienced an 85% increase in its urban population in the same period but the number of urban dwellers that had no access to safe drinking water doubled (*Ibid.*).

A major reason for poor access to water services in Sub-Saharan Africa is the inefficiency of water utilities, especially those serving urban areas. Many systems are characterized by high water losses, inadequate revenue to cover operating costs, dilapidated and poorly functioning infrastructure, a lack of investment, low billing and collection efficiency, chronic water shortages, failure to meet existing demands, low coverage (especially for the urban poor) and corruption (World Bank 2004).

Kenya is one of the few countries in the world where urban drinking water coverage from improved sources actually declined between 1990 and 2004 (WHO/Unicef 2006: 15). Its urban water supply situation can best be summarized as follows: '[the] water supply in Kenyan cities is highly inequitable. Over 50% of the urban poor, living in slums, have no access to safe drinking water and end up paying vastly more for municipal piped water' (UN-Habitat 2005: 5). The Water Services Regulatory Board provides a gloomier picture: 'In the low-income settlements where an even majority of the urban poor live, only 20 per cent of the population have access to safe water, exposing them to relatively high tariffs charged by water vendors' (WASREB 2008: 1). A study carried out in Nairobi, Mombasa and Kakamega in 2000 concluded that 'the current water supply situation [in these three urban centres] is dismal' (Gulyani *et al.* 2005: 27) and another study undertaken in Kenya's three largest cities – Nairobi, Mombasa and Kisumu – in 2006 showed similar results (Citizens Report Card 2007). A comparison of the 'poor' and the 'non-poor' revealed that there are distinct inequities in access to mains connections, with the poor reporting less access. The difference is particularly dramatic in Kisumu, where only 7% of the poor reported having access to mains connections, compared to 81% of the non-poor.

Supported by western donors, the Kenyan government has made various attempts to reform its water sector since the 1970s but the results have been unsatisfactory.[3] According to GTZ, the German donor organization,

[3] For an overview, see Owuor & Foeken (2009).

this was largely due to the fact that the water departments responsible were fully controlled by the municipal authorities. Hence the 1995 decision to privatize the sector, at least partially, by creating water and sewerage companies. Although these would still to be owned by the municipalities, they were also to be fully responsible for their own finances (Onjala 2002). This reform was formalized in the Water Act 2002.

Although from a global perspective the setting up of water and sewerage companies from above may not necessarily be considered innovative, this was certainly the case in the Kenyan context. This is dealt with in more detail in the next section. What then follows is a description of how this innovation was transformed in one low-income neighbourhood in Kisumu in two major steps: (i) by creating a local water supply system that was developed, constructed and managed by the community itself; and (ii) by transforming this project into an officially registered water and sewerage company at neighbourhood level (i.e. from below).

Kenya's 2002 water reforms

Even in the 1980s as part of the structural adjustment ideology advanced by the World Bank and the International Monetary Fund (IMF), privatization was seen as the best way of getting the water sector in developing countries on its feet again.[4] The most radical form is full privatization (divesture), when a private company becomes the owner of the infrastructure and takes full responsibility for operations, maintenance and investment. The government's role is thus reduced to one of regulation.[5] Much more common though are various forms of public-private partnerships (see Table 7.1). Usually a distinction is made between 'service contract', 'management contract', 'lease contract', 'concession contract' and the so-called BOT type contract[6] (Budds & McGranahan 2003; World Bank 2004; K'Akumu 2006; Kirkpatrick *et al.* 2006). Responsibility for service provision is shared between the public and the private sector, with differing

[4] See Prasad (2006) and Wambua (2004) for an overview of the privatization debate in general, and in Kenya in particular.

[5] This model is not common and has only been adopted in England and Wales (Budds & McGranahan 2003). Nowadays full privatization as a way of reforming the water sector is seen by many as undesirable and unnecessary (Hukka & Katko 2003).

[6] BOT stands for Build-Operate-Transfer, i.e. the private company is authorized to build and operate the supply system but will transfer it to the public sector after a certain period of time. Variations are the ROT (Rehabilitate-Operate-Transfer) and the BOO (Build-Own-Operate) systems. See also Akintoye & Hardcastle (2004) and Otiego (200).

levels of responsibility being delegated to the private partner depending on the type of contract.

As in other countries in Sub-Saharan Africa, achieving Kenya's socio-economic development goals is highly dependent on the availability of a good-quality and adequate water supply. The government's long-term objective is to ensure that all Kenyans have access to clean potable water and that water is available for key economic activities too (MWI 2005; Kenya 2006b). The water-sector reforms currently being implemented under the Water Act 2002 of the Laws of Kenya are designed to contribute to the realization of its long-term objective (Kenya 2002). In the Act, autonomous water and sanitation (or sewerage) companies – so-called WASCOs – are given responsibility for providing water and sanitation services within urban areas. In other words, they are the direct Water Service Providers, with the lead partners in such ventures normally being the local authorities. WASCOs operate within the jurisdiction of the Water Services Boards that are instrumental in their registration, incorporation and monitoring. The Act requires that a Water Services Board draws up a contract with a Water Service Provider in a Service Provision Agreement but the Water Services Board remains the legal owner of water and sewerage assets in their areas of jurisdiction (WASREB 2008: 2). In terms of the public-private partnerships presented in Table 7.1, the concession type of contract thus seems the most applicable in the Kenyan situation.

Table 7.1 Allocation of key responsibilities for private participation options in the water sector

	Service contract	Management contract	Lease	Concession	BOT type of contract	Divesture
Asset ownership	public	public	public	public	public/ private	private
Capital investment	public	public	public	private	private	private
Commercial risk	public	public	shared	private	private	private
Operations/ maintenance	public/ private	private	private	private	private	private
Contract duration	1-2 years	3-5 years	8-15 years	25-30 years	20-30 years	indefinite

Source: Budds & McGranahan (2003: 89)

WASCOs have to be managed commercially, which includes signing performance contracts, cost recovery and sustainability within a context of efficiency, operational and financial autonomy, accountability and strategic (but minor) investments. The key word is not privatization but commercialization: water is considered by the Kenyan government as both a social and an economic good that should be available to all Kenyans at a price that reflects its market value (cost recovery). Put differently, water services have to be managed 'in accordance with sound business principles' (Section 57(5)(d) of the Water Act – Kenya 2002). As Wambua (2004: 7) argues, 'through commercialisation, the Water Act 2002 requires local authorities to form autonomous water and sewerage companies with independent Boards of Directors to provide water services and re-invest (ring-fence) water revenues in service delivery improvement'.

The government also recognizes that the poor cannot afford to pay high prices, a problem that has to be solved by offering subsidized rates. Sections 11(1) and 11(2) of the Act laid the foundation for the National Water Resources Management Strategy (NWRMS – 2006-2008) (Kenya 2006a).[7] Its overall goal is 'to eradicate poverty through the provision of potable water for human consumption and water for productive use' (*Ibid.*: 4). In short, WASCOs are supposed to improve access to water and sanitation services for poverty reduction and sustainable development. In fact, the core mandate of the WASCOs is to provide effective, efficient, adequate and safe water to all their customers.

The Kisumu Water and Sewerage Company (KIWASCO) in the western city of Kisumu on the shore of Lake Victoria provides services over an area of 297 km^2 to a population of almost 400,000 (Kenya f.c.). The company was set up in 2001, i.e. prior to the 2002 Water Act, and became operational in July 2003 as a limited company after the transformation of the Water and Sewerage Department of the Kisumu Municipal Council into KIWASCO. Compared to several other Kenyan WASCOs, the company's record so far has not been very impressive. In 2008, it produced about 18,000 m^3 of water daily but demand was estimated to be about 45,000 m^3 a day. The result has been severe water shortages. Moreover, Kisumu has one of the highest levels of unaccounted-for water in Kenya, i.e. water

[7] The NWRMS (2006-2008) is also based on three other policy papers: (i) Sessional Paper No. 1 of 1999 on National Policy on Water Resources Management and Development; (ii) the Economic Recovery for Wealth and Employment Creation Strategy (2003-2007); and (iii) the Poverty Reduction Strategy Paper (Kenya 2006a: 3).

that is provided but not paid for. When KIWASCO started its operations, the level of unaccounted-for water was 75% although this was reduced to about 62% by 2008. This (still) high level can be attributed to non-functional (static) meters, illegal connections, burst pipes and leakages.[8]

A large percentage of the municipality's population is concentrated in low-income settlements, most of which are not connected to the KIWASCO water supply network. This means that only a third of the total population is served by KIWASCO's water network. One low-income settlement that was not served was Wandiege and, in a bid to improve their water supply and sanitation situation, residents started a community water supply project, which is currently operating as a Water Service Provider like KIWASCO.

The Wandiege Community Water Supply Project: From inception to implementation[9]

Wandiege is part of the informal settlement known as Manyatta B on the east side of Kisumu town. Before the project, Manyatta B, a peri-urban, low-income, residential area, lacked a decent water supply system and sanitation facilities. Water used to be scarce in the area, with the only source being the nearby River Auji. Although a main KIWASCO water pipe passed near the area, it carried no water due to a lack of pressure. The only possibility for accessing KIWASCO water was to tap it from a source far away (0.5 km to a few km, depending on where people lived). This would have been a very expensive proposition for Manyatta B residents.

The sanitation situation was equally poor due to the local terrain, with its poor soils and water logging during the rainy season. Pit latrines in the area often experienced problems during periods of heavy rainfall and overflowed and/or collapsed due to their weak and shallow foundations. It was expensive for residents to build a deeper and reinforced pit latrine that could withstand the nature of the terrain.

The inception

The idea of a community water supply project in Wandiege was initiated in the fourth quarter of 2001 through a different programme at Wandiege

---

[8] Interview with the KIWASCO Financial Manager, 13 October 2008.

[9] Our thanks go to Mr Naum Mbeya Obondo (Chairman, Board of Directors, Wandiege Water and Sanitation Company Ltd) and Ms Hellen Omondi (Treasurer) and Ms Jane Akendo (Revenue Clerk) for this information.

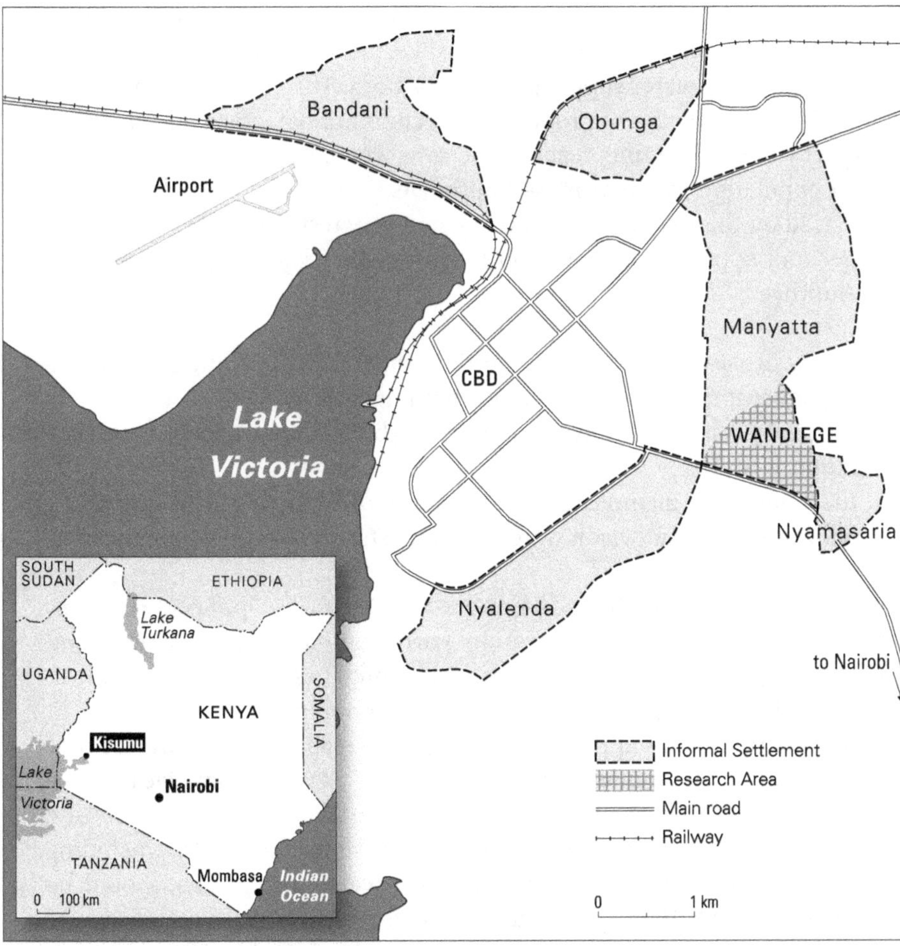

Map 7.1 Location of Wandiege in Kisumu

Primary School that was focusing on health and development issues in Africa. In the course of achieving its objectives, the programme noted that water and sanitation were major concerns in the area and a self-help group was established. This was initially to sensitize the community to the area's poor water and sanitation situation and the need for some sort of intervention. With the sensitization campaign, community support and technical advice from an NGO called SANA International, the idea of a community water supply project was born. The community, together with SANA, developed a proposal for a community water and sanitation project while community resource persons were identified to sensitize the local population to the proposed initiative.

Based on the vision of a society where every resident enjoys safe, clean water and appropriate sanitation facilities, the project's objectives were to improve:

- access to safe and clean water for the residents of Manyatta B by installing 500 water outlets/connections within 3 years
- the health status of Manyatta B residents by providing appropriate sanitation facilities
- the household incomes of the residents of Manyatta B by providing affordable, safe and clean water for domestic and commercial purposes.

The folder[10] also mentions the two strategies to be applied to accomplish these objectives: (i) community participation and involvement; and (ii) community mobilization and sensitization. These would ensure a feeling of full ownership of the project by the community.

This participatory approach led to the development of a project proposal. It was quite unique (certainly in Kenya) and very innovative because it involved not only the drilling of a borehole and the construction of a standpipe but also a whole water supply system for the locality, including storage tanks, pipelines, water kiosks and water meters, that would be independent of KIWASCO. In short, the community wanted to 'take water to their doorsteps'. Since the borehole was to be drilled on Wandiege Primary School's compound, the proposal also included the installation of an electricity power line to the school. Finally, the construction of a number of appropriate sanitation facilities (so-called Ecosan and Sanplat latrines) was also included.

[10] 'Wandiege Water and Sanitation Company Limited' (not dated).

The proposal was submitted to the French Embassy in Nairobi and to a Dutch NGO, Cordaid, for funding. The French Embassy contributed money to fund the drilling of the borehole and the construction of a tower for water storage tanks and an office. KSh 3.5 m was still needed at this point to construct the water pipelines and water kiosks, and Cordaid was willing to contribute KSh 2 m provided that the community raised another KSh 1 m itself. The local authorities and the central government pledged KSh 200,000 and KSh 300,000, respectively.[11]

Cordaid's insistence on the community's contribution was due to its conviction that such a community-based project would only succeed if the community (partly) owned it. This was a challenge because the community could not raise such a large amount of money immediately. To go about it, it initiated a recruitment drive of 1000 project members who would each contribute KSh 1000 as 'shareholders', but only about 300 shareholders were recruited. This unexpected shortfall was for various reasons. Firstly, some people felt that they should not have to pay anything since it was a donor-funded project. Secondly and despite a spirited sensitization campaign, others still showed ignorance and a lack of understanding about their role in the project and, thirdly, being a poor community, some could simply not afford the membership fee.

The community met again to look for other ways to go about this new challenge. They decided that 'membership' could also be through the provision of labour up to a perceived amount of KSh 1,000. Labour constituted, for instance, digging trenches for water pipes, laying pipes and constructing water kiosks. Another challenge in the initial stages was that the local authority and central government never honoured their promises. So the community sought funding from Kisumu Town East's Constituency Development Fund (CDF),[12] which gave the project KSh 500,000.

Implementation

Implementation started in 2001 with SANA International as the executing agency. The Ministry of Water and Irrigation, together with SANA, provided technical advice and the project was completed in 2006. It included a borehole to a depth of 110 m; a pumping station; a tower with two 10,000-litre storage tanks; a small office at the bottom of the tower; a pipeline

11 At the time, US$ 1 was equal to about KSh 62 and € 1 about KSh 80.

12 CDF is a (devolved) publicly funded institution that targets development projects at the constituency level.

Photo 7.1 The pumping station-cum-kiosk and storage tower in the school compound [Photo: Dick Foeken]

Photo 7.2 One of the water kiosks operated by the project [Photo: Dick Foeken]

system of about 5 km; seven water kiosks; and 60 connections to private houses.

The borehole, pumping station and tower were built on the compound of Wandiege Primary School (Photo 7.1) and in return, the project provides the school with free water as well as free electricity to one of the school blocks. The kiosks were located according to the service population and a needs assessment (Photo 7.2). Three are in school compounds.[13] As for sanitation facilities, 14 eco-sanitation (Ecosan) and 91 sand-platform (Sanplat) latrines had been constructed by October 2008.[14]

About a year after the completion of the project, the Wandiege Community Water Supply Project became an official WASCO – Wandiege Water and Sanitation Company Limited (or Wandiege WATSAN Ltd) – after the signing of a Service Provision Agreement with the Lake Victoria South Water Services Board. As such, it gained the same legal status as

13 In addition to Wandiege Primary School, these include St Brigit's Primary School and Nyamasaria Secondary School.

14 Ecosan is an ecological sanitation system that recycles the nutrients in human excreta and urine for use in agriculture. Sanplat is a 'sanitation platform' in the form of an improved latrine slab that makes simple latrines hygienic and safe.

its 'big brother' KIWASCO. The company is run on a non-profit basis, with any profits made being reinvested in the infrastructure or whatever is needed by way of further development. As a result of a sound financial policy (see below), the company has been able to extend its network system and, by September 2011, another 3 km of pipeline had been laid, 148 metered connections established[15] and a complete chlorine doser for treating the water had been installed. These connections serve a population of 10,000-15,000 residents.[16]

All the connections are metered but instead of placing meters at the end of each subsidiary pipe, i.e. a pipe branching off from the main pipe and leading to a connection, which is common in Kenya, each meter is mounted in a so-called meter chamber at the point where the subsidiary pipe leaves the main pipe. This has two important advantages: (i) it prevents other people from illegally tapping water from the subsidiary pipe because that would mean that the person with the legal connection would have to pay for the illegal connections as well; and (ii) any burst pipe or leakage in the subsidiary pipe will be reported to the company immediately because nobody will want to pay for water s/he has not received. This means that the water supply for connected households is very reliable. And for the company, the net result is that the percentage of unaccounted-for water is extremely low.[17] A concomitant advantage for the company is that meter reading is much easier because instead of having to go to each household, only the main pipelines have to be read.

Operations

The original water kiosks are operated and managed by the company and are together with the private connections the main source of income for the company. The kiosks are run by young people from the neighbourhood who are employed by the company, one of whom is also the pump operator. There is also a revenue clerk, a line patroller and a security officer who are all paid by the company. The privately-run water kiosks are

15 These connections include the original six water kiosks run by the company, 14 privately-run kiosks and two kiosks operated and managed by the church. The rest are individual connections to homes.

16 Figures obtained from the Chairman of the Board's presentation at the Annual General Meeting of the Wandiege Water and Sanitation Company, 8 June 2011.

17 Burst pipes, leakages and illegal connections are common in Kenya and lead to very high percentages of unaccounted-for water and hence huge losses for WASCOs. See the above-mentioned example of Kisumu/KIWASCO.

supposed to achieve one of the project's objectives of alleviating poverty in the community through income-generating activities. They are a water point and a place where people come to buy water, and some have a tank to store water for when there is no running water from the pipe.

According to the Chairman of the Board of Directors, connection and consumption charges are 'very affordable' and 'the lowest' in the area. Both tariffs are based on the project's primary principle of affordability and poverty alleviation, i.e. the tariff should be as affordable as possible while the members in the water business could make a 'socially acceptable' profit even if they employ staff to work in their business. In addition, the lower connection and consumption charges are a way of encouraging more connections to increase the company's revenue.

According to the Chairman of the Board of Directors, the connection and consumption charges are 'very affordable' and 'the lowest' in the area. Both tariffs are based on the project's primary principle of affordability and poverty alleviation. In addition, the lower connection and consumption charges are a way of encouraging more connections to increase the company's revenue.

In May 2011, connection costs went up by 22% for kiosks and by 40% for individuals. Consumption costs for individuals and institutions increased by KSh 5 per cubic metre, which means that the rise was higher as their consumption was lower, ranging from about 20% for 'low' consumers to 10% for 'high' consumers. Kiosks were charged about 26% more by the company. These price increases were inevitable because of increasing operational costs, such as wages, maintenance costs and especially the cost of electricity.

Table 7.2 Connection charges as of May 2011 (in KSh)[18]

Type of costs	Type of connection		
	Individual	Institution	Kiosk
Application	200	200	200
Deposit	1,500	7,000	12,000
Connection	1,250	3,500	2,500
Meter rent	100	175	125
Total	3,050	10,875	14,825

Source: Wandiege Company Office (September 2011)

[18] In September 2011, KSh 100 was equal to about € 1 or US$ 1.3.

Table 7.3 Charges for water consumed as of May 2011 (in KSh)

Amount consumed	Type of use		
	Individual	Institution	Kiosk
0-6 m^3	250*	35	
7-20 m^3	30	35	
21-40 m^3	35	40	
41-60 m^3	50	45	
Over 60 m^3	55	55	
0-180 m^3			35
181-250 m^3			40
Over 250 m^3			50

* Flat rate
Source: Wandiege Company office (September 2011)

The price of water for individual consumers from a water kiosk is set by the company. Originally this was KSh 1 for a 20-litre jerry can but the price increased to KSh 2 in May 2011. Compared to prices in, for instance, nearby Homa Bay town where a jerry can of water costs KSh 3 to KSh 5,[19] this is still relatively low. And despite the increase of 26% that the kiosk holders have had to pay to the company, selling water is still profitable. For instance, a kiosk holder in the highest consumption category pays the company KSh 50 for one cubic metre but receives twice this amount from his clients.

Figure 7.1 presents the financial results for the years 2007 to 2011. The first two years show a modest profit of KSh 67,000 and KSh 55,000 respectively. The following two years, the company did better, given the increase of 48% in 2009 over the previous year. As expenditure rose less quickly, the net profit grew more than six-fold. The net profit in 2010 was slightly less than in 2009 – because expenditure rose more than income – but increased again in 2011. On the whole, the figures in Figure 7.1 suggest a sound financial policy.[20]

19 Authors' own observations, September 2011.

20 Critics have, however, warned that these positive results would be totally insufficient to restore the whole system if it were to break down due to some major eventuality. According to them, the project's philosophy is still too much 'self-help' and 'non-profit' and too little 'business' (Information from Moses Mwangi, personal written communication).

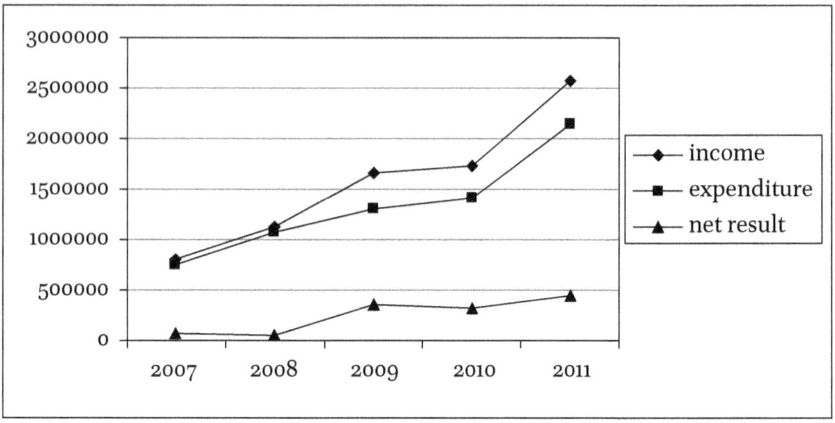

Source: Wandiege Company Office (September 2011 & May 2012)

Figure 7.1 Financial overview, 2007-2011 (in KSh)

Management

All paid-up and registered members of the project are shareholders in the Water Company. These individuals and institutions are represented by an elected Board of Directors (hereafter referred to as the Board). The eleven-member Board consists of a Chairman, Secretary, Treasurer, representatives from both SANA International and Wandiege Primary School and five other members. Its role is to represent the shareholders in all aspects of decision-making; negotiate on behalf of the Company; decide on policies; look after employees' welfare; and act as a Management Team (as none was previously in place). More often than not, the Board takes decisions on behalf of the shareholders, who are normally briefed at the Annual General Meeting on all the decisions taken on their behalf. However, if shareholders have to be consulted, a Special General Meeting is called.

Board members do not earn a salary and are totally committed to the welfare of the members and the community. However, they do receive KSh 500 for attending board meetings and are given lunch at official functions. This practice was started recently to show appreciation for the Board's commitment to the project.

The Wandiege community works with a number of task forces to coordinate issues of common interest with the community. Water and sanitation, solid-waste management and education are examples of existing task forces and the community has also formed a larger Wandiege Residents' Association. All of these are off-shoots of the water project.

The success of the Wandiege project

A success story, but with inherent challenges

Since its inception in September 2011, the Wandiege project has developed and expanded in various ways. Firstly, it has been transformed from a community water supply project into a water company in the category of community water service provider. Wandiege Water Company is now recognized by Lake Victoria South Water Services Board through a Service Provision Agreement as stipulated in the Water Act 2002. Secondly, the area it covers has expanded from its initial focus on Manyatta B sub-location to parts of a neighbouring sub-location (Kasule) where the demand for water is equally high. Thirdly, the number of connections has increased to almost 150 and there is still more demand for connections. With more connections, revenue more than doubled in the period between 2007 and 2010 and net profits increased almost five-fold. Fourthly, the company has been able to take on a permanent workforce of nine persons, install an automatic water pump and chlorine doser, and implement a simple but accountable and efficient meter-reading, billing, revenue-collection and record-keeping system. And finally, the provision of good sanitation facilities has resulted in a marked reduction in the number of cases of water-borne diseases reported in the area.[21]

During the same period, the company experienced a number of challenges. Firstly, putting water pipes through people's homes and acquiring land on which to erect water kiosks is difficult because land tenure in Wandiege is freehold. This has required sensitivity and lengthy negotiations. Secondly, there have been cases of theft and vandalism of pipelines and meters that have caused unaccounted-for water and replacement costs. The theft and vandalism are suspected to be the work of competitors (such as water vendors) who are not happy with the company's success and rapid expansion. Water vendors are apparently not seeing the high profits they used to enjoy when there was a lack of water before Wandiege's intervention. Thirdly, local politics has occasionally impacted negatively

[21] There are no official figures to confirm this but it can be deduced from the fieldwork findings of two Masters students in November 2009 (Chung 2011; Mutune 2012). Incidences of water-borne diseases in households in Wandiege were half the number of those in an area without any water supply intervention, namely Bandani, which is comparable to the situation in Wandiege before the project started. In addition, most of the respondents in Wandiege indicated that the number of cases of water-borne diseases had declined as a result of the project (Mutune 2012).

on the company's intended operations. For example, a local bank, which had agreed to provide the company with a loan in 2010, pulled out after it emerged that the government intended to take over the project. Some politicians argued that the project was donor-funded and was therefore a government project. Being a registered community water service provider recognized by Lake Victoria South Water Services Board seems to have resolved this misconception. Other challenges include frequent interruptions in power supplies, the occasional non-payment of water bills, and unhealthy suspicion and competition from KIWASCO, NGOs, other residents' associations, community development associations and self-help groups.

Despite the challenges the company has been facing, the project is generally seen as having been a success. This is, for instance, shown by the fact that it is acting as 'a learning centre for other organizations and students from higher-learning institutions, nationally and internationally'.[22]

The sustainability of the project

The company is working on further improvements and extensions to fully realize its objectives and vision of a society where every resident enjoys safe, clean water and appropriate sanitation facilities. For example, a piece of land has been acquired for the construction of a bigger office and a higher-capacity storage tank to ensure that there is always a constant water flow even if there is a power blackout or a drop in pressure. By September 2011, two-thirds of the land costs (totalling KSh 600,000) had been paid. Moreover, with the help of SANA International, a loan of almost KSh 7 m was obtained from Cordaid to extend the pipe system by another 3 km, construct a storage tank and five new meter chambers and install 55 new water meters.

There are various plans for the (near) future:

- The company intends to negotiate with KIWASCO for an alternative source of water through KIWASCO's Delegated Management Model.[23]

22 Quoted from the Chairman of the Board's presentation at the Annual General Meeting of the Wandiege Water and Sanitation Company, 8 June 2011.

23 The Delegated Management Model (DMM) was a KIWASCO initiative implemented in an informal settlement called Nyalenda. KIWASCO links an area to one of its main water pipelines (by means of one or more subsidiary pipelines) but management of the sub-system is in the hands of a Master Operator who is a representative of the community but paid by KIWASCO. This is supposed to create more responsibility among the

The advantage for Wandiege Company would be that if there is a serious problem with their water supply system, they could access KIWASCO water on a temporary basis.
- Since the system is dependent on the provision of external electricity, the company hopes to develop its own alternative source of power.
- The management structure will be transformed. The eleven-member Board currently takes all the decisions but the idea is to create a Management Team that would report to the Board. The company is also planning to engage semi-autonomous professional operators.
- Acquiring a computer to automate their management and office operations would allow bills to be paid directly into the company's bank account.

The future and sustainability of the project will depend to a large extent on how the community nurtures and enhances the factors that have contributed to its success, namely, strong community participation, the commitment of members, ownership of the project and the delicate balance between business principles and access to safe and affordable water. There is no doubt that these should be coupled with appropriate and sustainable sources of energy, alternative sources of water, innovative management, operations and maintenance, and continued partnership with KIWASCO.

Conclusions: Innovation and transformation

Referring to the public-private partnerships outlined in Table 7.1, the Wandiege project is not easy to position. The major difference between the categories in the table and the Wandiege case is that the former are created top-down and the latter bottom-up. In this sense, the Wandiege project can be seen as an innovative way of setting up a specific type of public-private partnership in the water and sanitation sector. Despite being a WASCO, Wandiege WATSAN does not, like KIWASCO, fit in the concession type of construction because it owns all its assets. Nor is it a true BOT type either because it does not intend any transfer to the public

community for their own water supply, while KIWASCO hopes to reduce the number of illegal connections, the high percentage of uncounted-for water and 'spaghettization' (large numbers of flexible pipes illegally connected to the company's system). See also World Bank (2009).

sector in the (near) future.[24] The various options shown in Table 7.1 all relate to existing urban-wide, public water providers that have entered into some kind of public-private partnership. The case of Wandiege does not appear in the table so in this sense it is innovative in itself. Legally, it is a private company that owns its assets, invests the capital, takes commercial risks and takes care of the operations and maintenance. Yet it is not a case of divesture because it did not take over something that already existed. Moreover, the project is very localized, which is different from the types of public-private partnership shown in Table 7.1.

While the Wandiege community water project can be regarded as having transformed a Kenyan innovation, the project is in fact quite unique and innovative in its own right. First of all, it is much more than a donor-funded borehole and standpipe, and involves a whole water supply infrastructure. Secondly, some technical innovations have been realized, one example being the meter chambers. Thirdly, on the one hand the management is a kind of collectivity while on the other the project contains an element of privatization because people buy water at market prices. However, solidarity can be found in the price differentiation and the way water is accessed (pump, individual connections) by various types of customers and the way of ensuring that more vulnerable groups also have access to water. Fourthly, despite the initial and then subsequent challenges, the project owes its success to a number of favourable factors: (i) the Water Act 2002 that supports such initiatives; (ii) community involvement at all stages of the project from conception to implementation; (ii) the project is owned, co-implemented and managed by the community, including committed shareholders and a Board of Directors; (iv) profits are reinvested to improve and expand the water supply to the community; and (v) the continued support and commitment from SANA International and other well-wishers.

The Wandiege project has resulted in several (other) transformations. The most obvious is the improvement in water provision to the local people. Secondly, it has created employment for young people from the neighbourhood. Thirdly, the project has strengthened community feeling in the area in social terms, which started with the first sensitization campaign and culminated in the creation of several task forces and a residents' association. Fourthly, it created empowerment in the area due to capacity

---

24 A possible deal for the company with KIWASCO would be an example of a BOT type of public-private partnership.

building and the community-based ownership and management of the project/company. And finally, a sense of pride has developed: people see it as 'their project', a project that is recognized as an official company and visited by large numbers of interested people, even from abroad.

Whether the project and the ensuing company are a solution to the enormous water (and sanitation) problems that exist in Africa is difficult to say. It is doubtful whether such a project would succeed on a much larger scale. Only on a relatively localized scale can a community be effectively involved and have a feeling of ownership. Moreover, for a project like this to succeed, it needs at least one more favourable condition, namely a group of people with dedication, patience, perseverance, diplomatic skills and persuasive qualities, but there is no reason to believe that such groups could not be created elsewhere too. And even among its critics, there is broad consensus that the Wandiege Water Supply Project has not only contributed to the well-being of the local community but that it also showcases 'the country in terms of improving access to water through community involvement'.[25]

Postscript

On a short visit to Wandiege in May 2012 it was noted that some of the above-mentioned plans for the future had already been realized and/or were under construction. Four new community kiosks had been constructed; the tract of land for the construction of a new water reservoir and a bigger office had been bought with the company's own money (for KSh 600,000); and the new water tower with a 75,000-litre capacity had been built and connected to the KIWASCO water supply system (Photo 7.3). KIWASCO water flows twice a week (on Wednesdays and Saturdays) to fill the tank and is then distributed to the Wandiege customers. Although the tank is big, it empties within two hours 'because there is a high demand for water'.[26] The KIWASCO pipeline is connected with the Wandiege system at a junction, which is an improvement because the quality of KIWASCO water is somewhat better as Wandiege water can be a bit saline: 'people drink it, but not for tea';[27] and if one of the systems does not function, the other can serve as a fall-back option. The price people pay for KIWASCO water is the same as for Wandiege water: KSh

[25] Moses Mwangi, personal written communication.
[26] Mr Naum Mbeya Obondo, personal communication, 22 May 2012.
[27] *Ibid.*

Photo 7.3 The new sanitation block-cum-office and the new water tower [Photo: Dick Foeken]

Photo 7.4 A meter chamber in the school compound [Photo: Dick Foeken]

2 per 20 litres. The company negotiated a reduced price with KIWASCO to ensure some profit.

A two-storey building was under construction on the same plot (Photo 7.3). A sanitation block is being constructed with two separate sections (one for men and one for women), each with four toilets and three showers. Urine and faeces are collected in a 'dome' underneath, where biogas will be produced. The gas will be used for cooking. At first it will be available around the building but small pipes are possibly be laid in the future to houses at up to a maximum of 100 m from the installation. The whole system depends on the number of people who use the sanitation facility and, to attract more people, the idea is that a small market could be created on the company's plot. Both the water tower and the building have been paid for with a loan from Cordaid (through SANA International).

Finally, on the negative side, a number of the meter chambers (Photo 7.4) have been stolen. Since the people responsible have not been traced, the company was forced to revert to the old system and install meters at the end of the line in the compounds of connected households. Only those chambers that were secured (like the two in the school compound) were able to remain.

References

Akintoye, A.M. & C. Hardcastle (2004) *Public-private Partnerships: Managing Risks and Opportunities*. Armstrong: Blackwell Science.

Budds, J. & G. McGranahan (2003) 'Are the Debates on Water Privatization Missing the Point? Experiences from Africa, Asia and Latin America', *Environment & Urbanization* 15(2): 87-113.

Chung, H.C. (2011) 'Access to Water and Sanitation in the Informal Settlements of Kisumu, Kenya. Masters Thesis, Radbout University, Nijmegen.

Citizens Report Card (2007) *Citizens Report Card on Urban Water, Sanitation and Solid Waste Services in Kenya: Summary of Results from Nairobi, Kisumu and Mombasa*. Nairobi: CRC Initiative.

Gulyani, S., D. Talukdar & R.M. Kariuki (2005) 'Water for the Urban Poor: Water Markets, Household Demand, and Service Preferences in Kenya', Water Supply & Sanitation Sector Board, Discussion Paper No. 5, Washington, DC: The World Bank.

Hukka, J.J. & T.S. Katko (2003) *Water Privatisation Revisited. Panacea or Pancake?* Delft: IRC International Water and Sanitation Centre.

K'Akumu, O.A. (2006) 'Privatization Model for Water Enterprise in Kenya', *Water Policy* 8(2006): 539-557.

Kenya, Government of (2006a) *The National Water Resources Management Strategy (NWRMS) (2006-2008)*, Nairobi: Ministry of Water and Irrigation.

—— (2006b) 'Kenya National Water Development Report: A Report Prepared for the UN World Water Development Report II', Nairobi: Ministry of Water and Irrigation.

—— (forthcoming), *Kenya 2009 Population and Housing Census: Analytical Report, Vol. VIII: Urbanization*.

Kirkpatrick, C., D. Parker & Y. Zhang (2006) 'An Empirical Analysis of State and Private-Sector Provision of Water Services in Africa', *The World Bank Economic Review* 20(1): 143-163.

Mutune, T.N. (2012) 'The Impact of Interventions on the Urban Poor: An Analysis of the Wandiege Community Water Supply Project in Kisumu', Masters Thesis, University of Nairobi.

MWI (2005) *Strategic Plan of the Ministry (2005-2009)*, Nairobi: Ministry of Water and Irrigation.

Onjala, J.O. (2002) 'Managing Water Scarcity in Kenya: Industrial Response to Tariffs and Regulatory Enforcement', PhD Thesis, Roskilde: Roskilde University.

Otiego, L.A. (2006) 'Commercialisation of Water Services in Informal Settlements: Case Study of Manyatta, Municipality of Kisumu, Kenya', Masters Thesis, University of Nairobi.

Owuor, S.O. & D. Foeken (2009) 'Water Reforms and Interventions in Urban Kenya. Institutional Set-up, Emerging Impact and Challenges', ASC Working Paper No. 83, Leiden: African Studies Centre.

Prasad, N. (2006) 'Privatisation Results: Private Sector Participation in Water Services after 15 Years', *Development Policy Review* 24(6): 669-692.

UN-Habitat (2005) 'Kenya Urban Sector Profile', Nairobi: United Nations Human Settlements Programme (UN-Habitat).

Wambua, S. (2004) 'Water Privatization in Kenya', Global Issue Paper No. 8, Berlin: Heinrich Böll Foundation.

WASREB (2008) 'Impact: A Performance Report of Kenya's Water Services Sub-sector', Issue No. 1, Nairobi: Water Services Regulatory Board.

WHO/Unicef (2006) *Meeting the MDG Drinking Water and Sanitation Target. The Urban and Rural Challenge of the Decade*, Geneva: World Health Organization.

World Bank (2004) 'Reforming Infrastructure: Privatization, Regulation, and Competition', World Bank Policy Research Report, Washington DC: World Bank.

—— (2009) 'Improving Water Utility Services through Delegated Management', Washington: World Bank, Water and Sanitation Program – Africa.

8

'It is time to start my own farm': The unforeseen effects of two waves of resettlement on household formation in Zimbabwe

Marleen Dekker & Bill Kinsey

The newly formed Zimbabwean government embarked on a resettlement programme in 1980. With financial support from the UK and other donors, beneficiary families were allocated 12 acres of arable land. What was innovative in the Zimbabwean approach to resettlement was its focus on nuclear-family farms and its ban on the subdivision of land. Over time, many of the original settlers' sons married and wanted to start their own farms. Since legal restrictions prohibited de jure *subdivision of land and it was not easy for them to access land elsewhere, many adult sons remained co-resident with their parents and ploughed part of their parents' plot to sustain their own families. Albeit unintentionally, they were defying the approach being dictated by the new Zimbabwean government and were changing the characteristics of the families living in the resettlement schemes from nuclear households to vertically extended families. As a consequence, population pressure on available land resources increased substantially and threatened the initial success of the resettlement programme. This chapter documents these unintended developments and explores alternative routes to accessing land that became more common after 2000 when the Fast Track Land Reform Programme (FTLRP) offered access to new farms.*

Introduction

There is nothing new about changes in land law and large-scale population relocation in Zimbabwe. From the early 1900s until the 1970s, Africa's population experienced successive involuntary dislocations in order to vacate land for successive waves of colonial immigration and commercial farm settlement. Land laws were repeatedly amended to legalize these alienations. Small-scale irrigation schemes provided limited scope for voluntary resettlement and, following the advent of freehold title, there were legislated changes that allowed Africans to purchase land in certain (restricted) areas.

With independence in 1980 however, land reform and resettlement were introduced in Zimbabwe with a sense of urgency (Kinsey 1982) as there were serious concerns about dealing with the large numbers of refugees returning from Mozambique, Zambia, Botswana, Tanzania and elsewhere. These concerns had been fuelled by the registers compiled by the UN High Commission for Refugees of those who had indicated they would need land when they returned to Zimbabwe. Plans for resettlement were thus underway well before independence. The first of the beneficiary families were resettled six months after independence on farms that were purchased on a willing-seller, willing-buyer basis. The British Government agreed to meet half of the costs of resettlement, including purchasing the land and developing the necessary infrastructure in the schemes, provided the Zimbabwean Government matched the amount (Palmer 1990).[1]

This first phase of resettlement[2] after independence had been designed by agricultural technocrats of the old Rhodesian institutions based on previous population relocation experiences. Their designs were in the nature of blueprints and were intended to be implemented fairly mechanically and extremely rapidly. They also stressed compliance with a daunting set of rules and regulations. Although innovation was not primarily their concern, the lack of awareness among Rhodesian technocrats of resettlement experience elsewhere in the world ensured that some of the regulations imposed were unintentionally innovative. This was especially the case for those related to the abandonment of many of what had become traditional practices in the communal areas. For example, although there were scattered attempts to design resettlement models to cater for unmarried young people, more than 90% of the land-reform beneficiaries were

[1] The land-reform programme initially had four components: the establishment of family farms (Model A); collective farms (Model B); individual cultivation on a core estate (Model C); and extensive ranching (Model D). This chapter discusses Model A only.

[2] When characterizing resettlement following 1980, it has been common to identify multiple 'phases' depending on the emphasis and approach. One government official, ignoring the declared broad objectives, characterized this earliest phase simply as 'rehabilitative' (Chatora 2003).

resettled under a model that accepted only nuclear families rather than extended households. In their new homes, each family was allocated an identical amount of arable land, namely five hectares, and about half a hectare of residential land, and were forced to live in nucleated villages, sometimes up to 2 km from their farming plots. This was in sharp contrast to the pattern in their communities of origin where extended-family homesteads were situated alone in the middle of cropped land. Grazing land was also designated and shared communally. A series of permits regulated the utilization of each category of land and stipulated that a beneficiary could lose his/her land allocation by infringing any of the regulations. For example, certain conservation practices had to be followed, livestock numbers were limited and, importantly, the selling, renting, leasing, pledging or subdividing of any land was forbidden. Yet despite the fact that the period planned for arriving at full economic maturity for the resettlement schemes was 18 years, the planners were operating within a single-generation framework and did not cater for the land needs of the second generation. This chapter reveals what this meant for the long-term viability of the innovations.

Although there was nothing especially innovative about Zimbabwe's post-independence resettlement programme in a global context, in the particular context of a state emerging from an isolated and racialized past, both the scale and much of the content of the programme were novel. However the authoritarian and top-down mode of implementation that characterized earlier development efforts allowed for an easy transition to the post-independence era.

This chapter argues that two of the innovative elements – the focus on nuclear families and the ban on subdividing land – initially determined the shape and composition of the resettled households. The inexorable increase in the man:land ratio[3] over time, however, led to growing informal evasion of the rigid aspects of the innovative policies. It was as if necessity had become the mother of invention, leading to widespread counter-innovative behaviour. Today, thirty years later and amidst severe economic recession, these early resettlement schemes are inhabited by households very similar to those living in their original communities of origin.

[3] Man/land ratio is calculated as the number of household members divided by the acres of land planted with crops.

Following this introduction in which some of the characteristics of the land-reform programme of the early 1980s are explained, the aspirations of the farmers who resettled are described, and the increasing population and the processes of migration, especially after 2000, are highlighted. The data presented in this chapter are based on information from a household panel study in three resettlement schemes, the Zimbabwe Rural Household Dynamics Survey (ZRHDS), and the life histories of settler families.[4] The locations of the resettlement schemes are shown in Map 8.1.

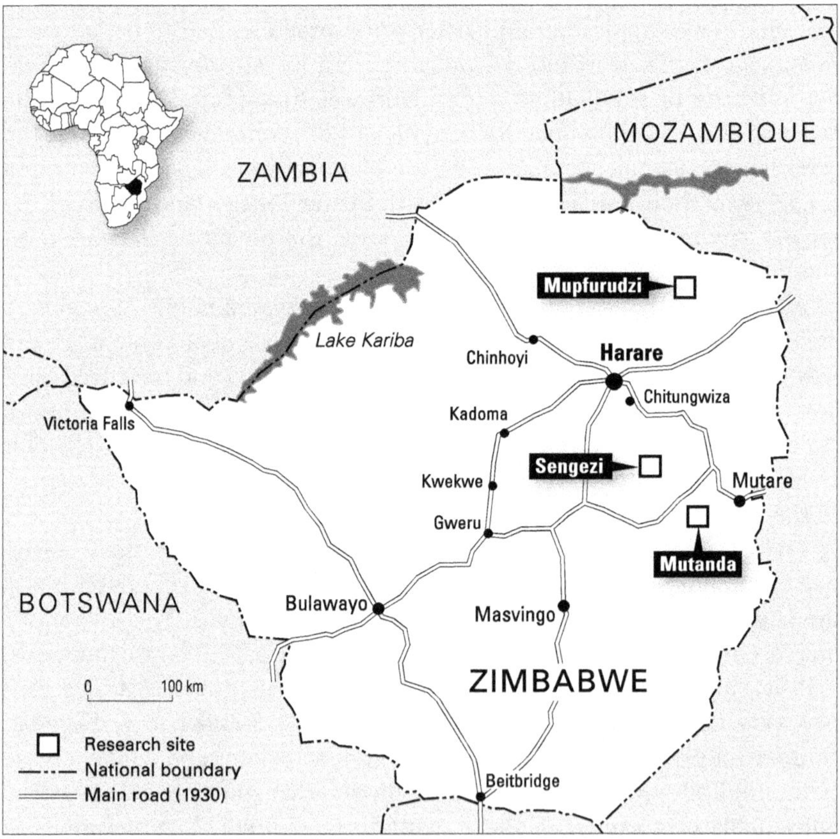

Map 8.1 Locations of the resettlement schemes

---

[4] The ZRHDS data are described in more detail in Kinsey *et al.* (1998).

Land reform in the early 1980s

The resettlement programme was based on voluntary resettlement and, in most cases, farmers' motivations for applying were self-interest. The applicants were screened according to several criteria: they had to be aged between 25 and 55, married or widowed,[5] and not in formal employment. Additionally, they had to be either a refugee or to have been displaced by the war (including those who lived in a protected village), landless or nearly landless residents of communal areas or small-scale farmers with insufficient land to maintain themselves and their dependants (Deininger *et al.* 2004). On resettlement, the settlers had to give up all rights to land in communal areas and were forbidden to take on work outside their new farm.[6]

In the early years after independence, the government invested heavily in infrastructure and support services in the resettlement areas. At least one service centre was established in each scheme, with a depot for seeds and fertilizers, space to establish businesses like grocery stores and grinding mills, a health clinic, and offices and houses for agricultural extension and veterinary staff. With material and labour input from the settlers, the government established primary schools and dip tanks for cattle, and the settlers had preferential access to credit and agricultural extension and veterinary services (Deininger *et al.* 2004). In the first agricultural season following resettlement, the government ploughed one acre and provided inputs for each household so that they could grow maize. Later, the government also assisted some villages with small projects, such as providing wire to construct paddocks or seedlings to start gum-tree plantations. In the late 1980s, the settlers could obtain a housing loan from the government to improve the standard of their housing.

Notwithstanding the fact that post-independence settlers received strong support at the outset, they were the *target* of planning rather than active participants in designing their own new communities. A chief concern of the government at the time was to try to ensure that the over-ambitious resettlement targets were met. The easiest way of achieving this was to keep resettlement simple, with a uniform, standard approach, and

[5] Women were only eligible for resettlement if they were widowed and had dependent children.

[6] These regulations were part of the mindset of the old Rhodesian technocracy yet were incorporated as regulations under the post-independence regime.

Photo 8.1 Many resettlement farmers took up cash crop production. Farmers handling their cotton harvest in Mupfurudzi Resettlement Scheme [Photo: Marleen Dekker]

shroud it in rules and regulations (Kinsey 1983). These regulations were to be enforced by the presence of resident government officials, the most important being the resettlement officer. Legally, resettlement officers had the right to withdraw permits and evict farmers who did not abide by the rules.[7]

The individuals holding the post of resettlement officer, as well as the agricultural extension officers, regularly exhorted those resettled to be agriculturally productive, threatening them with eviction from their resettlement areas if they failed to produce. Yet the standard five-hectare resettled holdings were never in reality designed to be a new, small-scale class of commercial farms. The planned income targets for the new holdings – after the eighteen years required to reach full economic maturity – were only intended to equate with the minimum industrial wage in urban areas. Thus an implicit objective was to keep the settlers on the land while denying them any incentive to join the flood of job seekers moving to the urban areas. The incentive of parity in income was judged to be insufficient and among the many rules the new settlers had to agree to at the outset was one prohibiting any off-farm employment.[8]

[7] The mere threat and occasional implementation of such evictions were reported to be sufficient to make most settlers abide by the rules.

[8] In the early 1990s, restrictions on labour migration were lifted, in part to allow resettled farmers to earn money due to the food shortages caused by the 1992 drought but also in response to an increasing number of defaults on agricultural loans (Kinsey *et al.* 1998).

Initially, the new settlement communities were disenfranchised and not able to vote for any form of political leadership beyond the purely local level. This anomaly was eventually rectified but the possibilities for strong local self-government were severely weakened in the late 1990s by the imposition of semi-hereditary traditional leadership in the resettled areas, in places where such leadership had been completely absent for more than 50 years. Thus the first generation of land-reform farmers in Zimbabwe spent more than two decades in a top-down, authoritarian administrative system that effectively discouraged any form of grassroots managerial innovation but provided an attractive resource-base for the farmers. The twelve acres of cropping land available in the resettlement areas was, on average, twice the size of land holdings in communal areas and was situated in areas with better soil quality and generally more reliable rainfall compared to the areas the setters originated from.

Aspirations of the new settlers

Most farmers settled in the newly established resettlement schemes between 1980 and 1982, mostly arriving with nuclear-family members and averaging a total of 7-8 persons. Not all household members moved to the new villages at the same time. Sometimes spouses remained behind in their previous place of residence with their children until their husbands had established a homestead and fields. And sometimes other children only joined their parents when a school had been built within a reasonable distance of their new homestead (Dekker 2004).

The settlers had clear aspirations for their future: some were modest, others more ambitious. Some mentioned that they just wanted to produce enough food, to be able to educate their children or to have sufficient money so they did not become indebted. Others had visions of the investments they wanted to make, first and foremost in cattle but also in housing and farming equipment, with some even mentioning tractors. A few aimed to buy a car or start a business such as a grinding mill or a shop. Taken together, they formulated their aspirations almost entirely in terms of farming and only a very few had dreams of diversifying into business. Land-reform beneficiaries in the 1980s and 1990s agreed almost unanimously

Given the importance of rural-urban migration in the spread of HIV/AIDS, this initial ban on labour migration, combined with the relative isolation of many resettlement villages, may mean that the number of cases of HIV/AIDS is lower in the resettlement areas under study than in many other parts of the country (Serra & Kinsey 2000).

Photo 8.2 A grinding mill in Mupfurudzi Resettlement Scheme [Photo: Marleen Dekker]

that their lives were better as a result of the land reforms, and this was supported by several studies that documented improved agricultural production, asset accumulation and higher household incomes (Hoogeveen & Kinsey 2001; Gunning *et al.* 2000; Owens *et al.* 2003).

Interestingly, the settlers did not define their ideal future for their children in the same way. Some 15 years after most were allocated their land, less than 3% of the resettled farmers mentioned farming as their career of choice for their children. Most hoped their children would become civil servants in the education or health sectors, their sons predominantly as teachers, drivers or administrators, and their daughters as teachers or healthcare workers. A minority wanted a technical, medical or security (army/police) career for their sons, or an administrative position for their daughters. Some children, of course, have been able to follow the career path desired by their parents. Many have not though, either by choice or simply by necessity. This is partly due to the (perceived) limited opportunities to do so after the structural adjustment programmes in the early 1990s reduced employment opportunities across the board. These same children have also faced challenges in gaining access to land.

Where do sons find land?

It is important to explore how sons have managed to access land to start their own farms and relocate their families. Although not very common, there are accounts of sons finding land in the communal areas from which their fathers originated:

> Mr B's son married a girl from the village in 1996. Initially the young couple settled on the stand of John's father but after a few months John was able to find his own stand in Madziwa Communal Area, where both his parents and his parents-in-law were living before they were resettled. The plot was located in Chief Mushowani's area where his father's relatives were still living.

Others found land in communal areas where they were considered 'strangers':

> Mr M's daughter married Mr C's son in 1995. Initially they resided on the stand of Mr. C in Mupfurudzi, where they cultivated together with Mr and Mrs C. In 1998 they moved to Rushinga Communal Area, northeast of Mount Darwin close to the Mozambican border. The pressure on land became too great as three married sons resided on his plot and wanted to use a portion of his land. In Rushinga there was still enough virgin land compared to Bushu Communal Area where they came from.

Alternative routes to acquiring land also opened up in new resettlement schemes:

> Mr M settled with his older brother in 1980. He was still young and his brother sent him to school as their parents had both died. In 1991, he married the daughter of another plot holder in the village and after a few years of cultivating jointly, the young couple was able to acquire a plot in Principe, a nearby resettlement scheme that was established in 1994.

Still other sons sought to gain land by means that were, strictly speaking, illegal:

> Farai stayed with his parents after he married his first wife and had two children. However when he married his second wife in 1998, his father suggested that it was time to start his own farm. Subsequently, Farai and his two wives and the children from the first wife moved out and squatted in the village where his father lived. They built a homestead in the bush and cleared fields to start cultivating.

In general, sons of resettled farmers found it more difficult to access farming land than sons of communal farmers. In communal areas, all members of a patrilineage from that particular area are entitled to their own residential plot and agricultural fields after they reach a certain age and status. In the virilocal marriage system, a bride comes from another lineage and thus generally leaves her natal family and lives at the *musha* (homestead) of her parents-in-law for a few years before she and her husband get their own farm (Pararadza 2010). Such allocations are usually made by the village head and may involve the redistribution of land cultivated by other family members.

The differences between communal areas and resettlement areas are shown in Table 8.1, which illustrates the place of residence of the married sons of the households in the ZRHDS study. This information was collected in 2000 and represents all the married sons of the household heads who were interviewed. Three interesting observations can be made. First, the most frequently mentioned place of residence for married sons of resettled families is at the same homestead as their parents (40%) or Harare (21%). Second, the figures also demonstrate that it is not easy for these sons to find land elsewhere; some (11%) have found land in another rural area, some in another village in the same resettlement area (4%), and some in their father's former home area (2%) or the neighbouring resettlement area (0.4%). Another small group is living on another homestead in the same village (4%), for example because they resettled independently of their parents in the early 1980s or when they married a neighbour's daughter after resettlement. The third observation concerns the considerable (and statistically significant) differences between sons in resettled and communal households. Sons from communal households more frequently live in Harare (or another urban centre) or in the same area but at a separate homestead and only a small fraction of married sons in communal areas reside on their parents' plot (10%).

Table 8.1 Place of residence of married sons of plot holders (%)

	Resettlement areas (N=511)	Communal areas (N=160)
Same homestead as plot holder	39.5	10.0
Same village as plot holder but another homestead	4.1	23.1
Another village in the same RA/CA*	3.9	4.4
Neighbouring RA	0.4	0
Former home area	1.8	0.6
Another rural area	11.4	9.4
Commercial farm	3.1	1.3
Rural business centre	3.1	2.5
Harare	21.1	33.8
Other urban centre	9.6	13.8
Other	2.0	1.3

* RA = Resettlement area; CA = Communal area
Source: ZRHDS data (2000)

Increasing land pressure in the late 1990s

The constraints on outside options outlined above impacted on the innovative approach to resettlement. Although most settlers initially arrived with nuclear-family members, over time the composition and size of the households changed considerably. The evolution of household composition is illustrated in Figure 8.1, which shows the proportion of household members that are nuclear-family members. In 1984, some 95% of household members were in the nuclear category (parents and children) but this figure declined to 60% of household members in 2000. Departing from the original nuclear-family approach laid down in the early 1980s, most settlers were living intentionally or unintentionally in extended-family arrangements by the late 1990s. By 2000, the average household had between 9 and 10 members and less than 25% of plot holders were living only with nuclear-household members (Dekker 2004).[9]

As it was not easy for married sons of settlers to obtain their own land or find alternative livelihoods either within or outside the resettled areas,

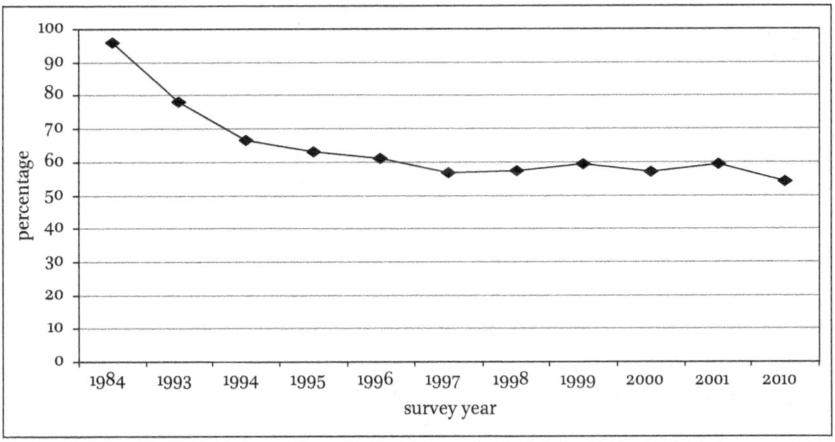

Source: ZRHDS data (1984-2010)

Figure 8.1 Percentage of household members belonging to the nuclear family

[9] Larger household sizes have been documented for other resettlement schemes as well (Harts-Broekhuis & Huisman 2001) and can broadly be explained by the labour requirements needed to cultivate the larger amount of land available to resettled farmers and the relative success of agricultural production, which has attracted additional household members (horizontal extension).

many remained living with their parent(s) and ploughed a portion of their parents' arable fields (vertical extension). This is *de facto* subdivision as the son(s) and his wife (or wives) cultivate part of the land by themselves and take full responsibility for investing in inputs and have ownership of their harvest. These crops are stored separately from the parents' crops and they have their own kitchen where they prepare and consume food. The mean number of kitchens on a plot was 1.6 for the full sample in 2000, with an average of 1.2 in communal areas and 1.8 in the resettlement areas. These mean figures hide a more diverse picture, with some 20% of the plots in resettlement areas having three or more kitchens, while this is true in only 6% of the households in communal areas.

Plot holder MM and his wife RM who live in one of the villages in Mupfurudzi Resettlement Scheme are a good example. MM settled in 1982 when he was 56, coming with his wife and seven children. His eldest son was 30 years old at the time of settlement and his youngest son was only nine. By 2000, their homestead had six kitchens;[10] all five sons were married (in 1984, 1988, 1991, 19.94 and 1997 respectively) and living with their wives on their parents' residential stand. With 11 grandchildren also living with them, the total number of residents was 23 persons.

The settlers themselves were increasingly aware of growing pressure on land, as is evident from comments made in the late 1990s. In a year with above-average rainfall, almost 25% of resettled plot holders indicated that their land was insufficient to support the family living on it. When the respondents were asked to consider their needs in five years' time, at least half of the farmers indicated their plot would not be sufficient to support their families. In reflecting on the reasons for this, only a small minority (5%) referred to physical constraints (rain and fertility) or a wish to diversify crops, while the large majority (95%) referred to the large (and still growing) families they needed to support and their children who would require land in the next stages of the demographic cycle. When probed on possible sources of land for their sons who wanted to farm on their own, nearly 56% of the plot holders in the resettlement schemes suggested they would share their land with their sons (Table 8.2). Roughly 25% of respondents claimed their sons would have to look elsewhere, preferably in government-initiated resettlement schemes. Only a few farmers expected new land to be cleared in their own community (6%) or access to it through the headman or council (3%).

[10] In Shona tradition, each wife in a polygamous household has her own kitchen.

Table 8.2 Farmers' perceptions as to where their sons would acquire land for farming

Category of farmers' responses	% of responses in this category
I'll share my existing plot with them; they have nowhere to go except this plot; I'll subdivide into sub-plots and allocate them; they'll take over my plot/my land	55.9
The government (or the party) will/should provide land; they will see what to do; they promised to provide land	18.3
They must find their own land elsewhere; there is no more resettlement	7.0
In this resettlement area, land will be set aside near the village or be obtained from neighbours, from those who leave or from those who are not utilizing their land fully. We'll clear more land in this village	6.0
The resettlement areas were established by the government, thus they should apply to the proper official (resettlement officer etc.)	4.6
In the communal areas: they must go to the headman to get their own land; they must return to the rural areas where we came from and apply to the chairman or the council	3.1
Not applicable because I have no sons; my sons are still young, or are away, or do not want to farm	3.1
They must get educated and find work elsewhere – in town, etc. No hard farming life for them	1.0
Nowhere: when the young grow up there will be no more land; all available land might go to squatters	0.7
I'll buy a plot for them	0.2

Source: ZRHDS (1995)

The growing pressure on land also becomes evident if the man:land ratio in the resettlement areas is considered. The ratio decreased from 1.19 per acre in 1997 to 1.0 per acre in 2001. At the same time, the man:land ratio increased from 0.65 to 0,77 in communal areas (Kinsey 2010). The increasing man:land ratio meant a reduction in *per capita* benefits from the first generation of land reform (Deininger *et al.* 2004) and potentially threatened food security.[11] What is important in this context, however, is also that it changed the nature of households in the resettlement schemes. In

[11] It is estimated that a minimum cropland area of 0.5 ha/person is required to provide an adequate diet (Lal 1989). The one acre available for a person in 2001 is below this norm.

2000, resettled households were still substantially different in composition compared to communal households. Ironically and in contrast to resettlement policy, the proportion of household membership in the nuclear category was then 10% higher in communal areas than in the resettlement areas (70% and 60% respectively). A disaggregation of household membership categories is presented in Table 8.3. Communal area households are predominantly composed of a head, a spouse and children, while resettlement area households host a higher proportion of sons and daughters-in-law and grandchildren[12] (as a result of *de facto* subdivision) and other blood or marriage relatives (horizontal extension).

Migration after FTLRP

Growing family sizes could have further eroded *per capita* revenue from farming and increased pressure on available land resources, eventually changing the nature of resettlement schemes and making them resemble communal areas. This scenario remains speculative however and has been

Table 8.3 Household members and their relationship to the household head in resettlement areas and communal areas (2000)

Relationship	Resettlement areas (%)	Communal areas (%)
Head	10.6	14.8
Spouse	9.3	12.0
Child	39.7	43.4
Parent/sibling	1.9	0.6
Grandchild	23.5	21.9
Niece/nephew/cousin	3.1	1.5
Son/daughter-in-law	6.5	3.7
Father/mother/brother/sister-in-law	0.8	0.0
Other relative	2.0	0.3
Worker/servant	0.2	0.6
Co-wife	0.5	0.0
Other	2.0	1.2

Source: ZRHDS data

12 As a result of the HIV/AIDS crisis, some grandchildren will be orphaned. There is however no reason to believe that the proportion of orphaned grandchildren is larger in resettlement areas compared to communal areas. The higher proportion of daughters-in-law in resettlement areas supports the argument that the next generation was not moving away from the resettlement areas (vertical extension).

overtaken by history.[13] Over the past decade, the resettlement areas have witnessed a large outflow of *de jure* landless adults who tried to obtain their own plots in the fast-track resettlement schemes that were implemented during and after the occupation of commercial farms beginning in 2000. This trend is illustrated by the following case:

> The son of Mr D married the daughter of Mr S in 1999. The young couple resided with Mr D and they were farming together. As the couple did not feel happy with this arrangement, they decided to build their own stand in the nearby commercial farms that were invaded in early 2001.

The opportunities to obtain land under the FTLRP (Fast Track Land Reform Programme), combined with the macro-economic developments that crippled agricultural production, employment and economic activities more generally, have had a profound influence on the mobility of people (Hammar *et al.* 2010). In 2010, we collected information on almost 3000 individuals that at one time lived in the homesteads of the respondents interviewed (Dekker & Kinsey 2011). Just over half of these individuals were no longer living with the respondents; the large majority (more than 85%) had left after 1999. This pattern is strongest in the areas with the highest agro-ecological potential, where many sons remained living and farming on their parents' plot after they got married and started their own families.

Looking more closely at the characteristics of the migrants suggests that it is the younger generation that has moved away over the past ten years. The average age of migrants was 26 but there is a clear difference in age between those leaving before 1999 (31 years of age) and after 1999 (25 years of age).[14] If we look more closely at the relationship of the migrant to the household head, it is clear that after 1999 the majority of individuals leaving are adult sons, daughters-in-law and grandchildren whereas 42% of the migrants who left before 1999 were adult children, their spouses and their children. After 1999, this figure was 62% of migrants. Migrants

---

[13] By the late 1990s, however, there was already clear evidence that while, in *per capita* terms, resettlement areas were clearly superior to communal areas in the value of crops grown, the land-reform beneficiaries were not necessarily better off in expenditure terms (Hoogeveen & Kinsey 2001).

[14] This is also evident if the demographics of the residents in 2000 are used to project the 2010 residents and the actual 2010 residents are compared with the projected residents. Especially in the 25-35 age category, a considerable number of men and women are 'missing'; they lived in the resettlement villages in 2000 but do not live there anymore. For women, the age category is defined more broadly (20-40 years) as they traditionally move out of their parental homestead when they get married.

often left as families, especially when seeking to take advantage of the opportunity to start an independent farm provided by the FTLRP.

The outmigration of (young) families to start their own farms is also evident if we look more closely at the area of destination: other rural areas (usually FTLRP schemes) were most often cited as the location to which migrants went. Interestingly, there is also a group reporting movements within the villages, clearly a new phenomenon. In Mupfurudzi, some village heads decided to create new residential homesteads in their villages to allow adult children and their families to occupy their own stands. These were delineated on the outskirts of the village on land that was originally designated as grazing area or public space. The agricultural fields belonging to the original household heads were not extended and kept being informally subdivided under the authority of the household head. This is clearly against the policy set out for the original resettlement schemes and can be interpreted as an (independent) act by the village heads in response to pressure from the villagers, especially those whose adult sons had not been able to acquire a new farm in the FTLRP.

Although household sizes have declined considerably since 2000, household composition in resettlement areas remained similar over the following ten years, with the proportion of nuclear-family members decreasing slightly from 60% in 2000 to 54% in 2010. At the same time, the nuclear-family arrangement in communal areas has become less common: nuclear families declined from 70% of the total in 2000 to 57% in 2010 with the proportion now being very similar to that in resettlement areas. Table 8.4 provides disaggregated figures on the composition of household membership by area in 2010 and illustrates the increasing resemblance of communal and resettlement households in terms of their composition.

Discussion

After independence in 1980, the newly formed Zimbabwean government embarked on a resettlement programme. In the context of Zimbabwean smallholder farming, the programme contained two innovative elements: a focus on nuclear-family farms and a prohibition on the subdivision of land. By reviewing the changing composition of households over time and the problems the second generation has had in gaining access to land, this chapter has documented a growing informal evasion of the rigid aspects of the innovative policies that led to widespread counter-innovative behaviour away from the nuclear-family arrangement to vertically extended households. It is far from clear whether the set of innovations embod-

Table 8.4 Household members and their relationship to the household head in resettlement areas and communal areas (2010)

Relationship	Resettlement areas (%)	Communal areas (%)
Head	14.6	16.3
Spouse	9.0	11.7
Child	30.7	28.6
Parent/sibling	2.3	4.6
Grandchild	20.6	22.5
Niece/nephew/cousin	12.6	9.1
Son/daughter-in-law	4.1	2.6
Father/mother/brother/sister-in-law	1.2	0.5
Other relative	1.9	1.0
Worker/servant	0.8	1.5
Co-wife	0.4	0.0
Other	1.9	1.5

Source: ZRHDS data

ied in the original resettlement policies and programme was opposed intentionally or coincidentally. The entire programme was so bound by cumbersome administrative procedures and elaborate rules and regulations – and so remote from the intended beneficiaries – that some breaches of the prescribed conduct were almost inevitable. As noted however, over time necessity became the mother of invention, and the second generation increasingly did what any far-sighted planner should have been able to foresee, namely they voted with their feet. They either demarcated their own domains within the original resettlement plots, thus effectively subdividing them, and/or they moved to new land as soon as it became available. But it remains debateable whether these actions constitute what we might consider systematic counter-innovation tactics. They would seem to be a response born of incredibly straitened circumstances and realistic optional responses are very difficult to imagine.

What can be said with certainty is that the evidence examined here provides further illumination of what Sikor & Müller (2009: 1307) call the limits of state-led land reform: 'land reform programs conceived by national governments in a top-down fashion and implemented by their administrative branches through bureaucratic modalities'. As Sikor & Müller (*Ibid.*) note, such approaches are incapable of accommodating diverse local political-economic contexts. Similar observations were made by Michael Cernea (1994), who noted that government agencies responsible for resettlement pass the organizational burden of executing relocation to unequipped and understaffed low-level bureaucracies. Thus poor policy-making is compounded by poor execution and there is almost no scope for rectifying policy oversights (Kinsey 2010).

Our analysis here also demonstrates once again that the direct benefits of land reform and resettlement are restricted to a single generation. We do not discount the powerful indirect benefits for following generation(s) – perhaps better health, better education and a wider range of opportunities – and some of these can be observed among the households we have been studying for so many years. But we are also observing the consequences of innovation bound in a straitjacket. Innovation is perhaps now more than ever needed in the realm of intelligent policy-making in areas relating to agricultural land use and the relationship between the rural sector and the rest of the economy. Mechanisms capable of detecting the unintended consequences of policy innovations well before they force administrators to rush their responses in a disruptive and wasteful fashion are particularly important.

References

Cernea, M.M. (1994) 'Social Science Research and the Crafting of Policy on Population Resettlement', *Knowledge & Policy* 6(3/4): 176-201.

Chatora, N. (2003) 'Resettlement and Beneficiary Support: Settlement and Resettlement Models in Zimbabwe'. In: M. Roth & F. Gonese, eds, *Delivering Land and Securing Rural Livelihoods: Post-Independence Land Reform and Resettlement in Zimbabwe.* Madison, Wisconsin/Harare: Centre for Applied Social Sciences, University of Zimbabwe/Land Tenure Center, University of Wisconsin-Madison, pp. 269-272.

Deininger, K., H. Hoogeveen & B. Kinsey (2004) 'Economic Benefits and Costs of Land Redistribution in Zimbabwe in the Early 1980s', *World Development* 32(10): 1697-1709.

Dekker, M. (2004) *Risk, Resettlement and Relations: Social Security in Rural Zimbabwe.* Tinbergen Institute Research Series No. 331. Amsterdam: Thela Thesis.

Dekker, M. & B. Kinsey (2011) 'Contextualizing Zimbabwe's Land Reform: Long-Term Observations from the First Generation', *Journal of Peasant Studies* 38(5): 995-1019.

Hammar, A., J. McGregor & L. Landau (2010) 'Introduction. Displacing Zimbabwe: Crisis and Construction in Southern Africa', *Journal of Southern African Studies* 36(2): 263-283.

Harts-Broekhuis, A. & H. Huisman (2001) 'Resettlement Revisited: Land Reform Results in Resource-Poor Regions in Zimbabwe', *Geoforum* 32(3): 285-398.

Hoogeveen, H. & B.H. Kinsey (2001) 'Land Reform, Growth and Equity: Emerging Evidence from Zimbabwe's Resettlement Programme: A Sequel', *Journal of Southern African Studies* 27(1): 127-136.

Kinsey, B.H. (1982) 'Forever Gained: Resettlement and Land Policy in the Context of National Development in Zimbabwe'. *Africa* 52(3): 92-113.

—— (1983) 'Emerging Policy Issues in Zimbabwe's Land Resettlement Programmes', *Development Policy Review* 1(1): 163-196.

—— (2010) 'Who Went Where . . . and Why: Patterns and Consequences of Displacement in Rural Zimbabwe after February 2000', *Journal of Southern African Studies* 36(2): 339-360.

Kinsey, B., K. Burger & J.W. Gunning (1998) 'Coping with Drought in Zimbabwe: Survey Evidence on Responses of Rural Households To Risk', *World Development* 26(1): 89-110.

Lal, R. (1989) 'Land Degradation and its Impact on Food and Other Resources'. In: D. Pimentel & C.W. Hall, eds, *Food and Natural Resources*. San Diego, CA: Academic Press, pp. 58-140.

Owens, T., J. Hoddinott & B. Kinsey (2003) 'The Impact of Agricultural Extension on Farm Production in Resettlement Areas of Zimbabwe', *Economic Development and Cultural Change* 51(2): 337-357.

Palmer, R. (1990) 'Land Reform in Zimbabwe, 1980-1990', *African Affairs* 89: 163-181.

Paradza, G.G. (2010) *Single Women, Land and Livelihood Vulnerability in a Communal Area in Zimbabwe*. African Women Leaders in Agriculture and the Environment. AWLAE Series No. 9. Wageningen: Wageningen Academic Publishers.

Serra, D. & B.H. Kinsey (2000) 'Child Fostering in Rural Zimbabwe'. Paper presented at the Opportunities in Africa: Micro-evidence on Firms and Households Conference. Oxford: Centre for the Study of African Economies.

Sikor, T. & D. Müller (2009) 'The Limits of State-Led Land Reform: An Introduction', *World Development* 37(8): 1307-1316.

9

'Cassava is our chief': Negotiating identity, markets and the state through cassava in Mwinilunga, Zambia

Iva Peša

After cassava was introduced into Central Africa, the crop gained such prominence in the course of the twentieth century that it was adopted as a staple food in areas such as Mwinilunga in northwest Zambia. Its arrival led to changes in cropping patterns, culinary practices and regimes of labour but also had repercussions beyond the agro-ecological and domestic spheres. This chapter examines how cassava transformed aspects of politics, society and culture and how the societal context simultaneously transformed cassava cultivation, its preparation and its use. To understand the process of mutual interaction between cassava and its environment, aspects of marketing, state policies and identity are considered. The crop was used as a tool to express group identity in Mwinilunga as well as to seek (or avoid) engagement with markets and the state. Cassava was imbued with multiple layers of meaning and came to be regarded as the chief among crops.

Introduction

A female British missionary at the beginning of the twentieth century described cassava *nshima* as a 'big, grayish ball with a consistency of African rubber and smelling like a mass of decayed vegetables in the garbage can' (Springer 1909: 167-168). In spite of its unappealing colour, texture and smell, this cassava-based meal became the staple food in Mwinilunga District, a part of Zambia's North-Western Province, at the expense of

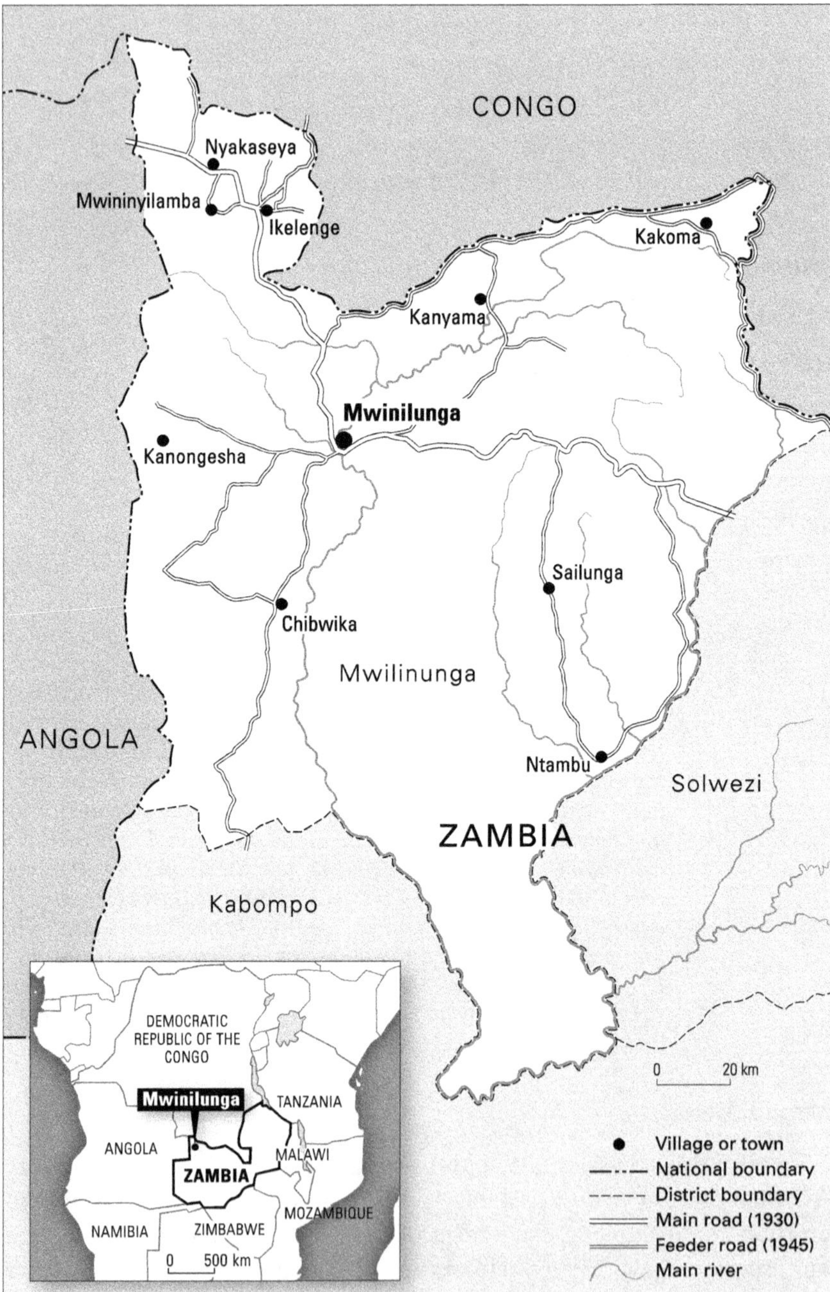

Map 9.1 Mwinilunga District

sorghum and millet, the long-established grain crops in the area.[1] Why did individuals overwhelmingly choose to grow and eat cassava to such an extent that the crop became an integral part of Lunda identity in spite of discouraging state policies and potential drawbacks? This chapter argues that cassava was used as a tool to assert autonomy by negotiating involvement with regional and national power constellations, especially markets and the state, on terms favourable to cultivators. Its distinct benefits meant that cassava, despite being a new arrival in the area, managed to shape intimate aspects of identity, as well as to unleash the potential for agricultural production in the area.

Africanizing a foreign introduction

As part of the 'Columbian exchange', the introduction of cassava to the African continent was inextricably linked to long-distance trade contacts (Crosby 1972). Originally a South American crop, cassava was brought to the Angolan coast by Portuguese traders in the sixteenth century (Jones 1959; von Oppen 1992; Vansina 1997) but its spread inland, which was spurred on by trade caravans, followed later. It is not likely that the crop was known in the Mwinilunga area before the end of the seventeenth century (Hoover 1978; von Oppen 1999). The introduction and spread of cassava did not constitute sudden events but were part of a slow and gradual shift that transformed existing agricultural practices to accommodate a foreign crop (Miller 1988). The acceptance, adoption and appropriation of this agricultural innovation took time to materialize but the effects of its introduction, which are irrefutably obvious today, extended over several centuries.

Far-reaching consequences have been attributed to the dissemination of cassava. It has been argued that the crop encouraged the incipient state formation of the Lunda polity, whose heartland lies to the north of Mwinilunga in present-day Congo.[2] As the grain crops previously cultivated in the area did not yield abundant harvests due to the poor Kalahari sands, population densities remained low. The consequently dispersed nature of the population complicated and discouraged forms of centralized rule.

---

1 The scientific names of the crops mentioned in this chapter are: cassava (*Manihot esculenta*); sorghum (*Sorghum bicolor*); finger millet (*Eleusine coracana*) and bulrush millet (*Pennisetum glaucum*); maize (*Zea mays*) and enset (*Ensete ventricosum*).

2 The court and capital city of the Lunda polity, Musumba, is on the Bushimaie-Nkalanyi River in today's Congolese province of Kasai-Oriental.

Cassava, however, could thrive in areas where other crops performed poorly and its high yields enabled more intensive settlement of the land. By facilitating higher population densities, which in turn permitted more direct forms of rule and control, the extension of the Lunda polity along the Upper Zambezi, where Mwinilunga lies, was achieved in the course of the eighteenth century (Schecter 1976; Hoover 1978; Papstein 1978). In present-day oral traditions, the connection between cassava and the forms of power exerted by the Lunda polity are firmly established, as the crop is said to have come from 'Lunda' with migrating chiefs rather than with Portuguese merchants from overseas.[3]

The role of the crop was limited at first as a radical reorganization of agricultural techniques was required to incorporate cassava cultivation. While sorghum and millet had been planted on ridges, cassava required the construction of mounds and its processing was onerous as the roots had to be dug up and soaked in water to remove any potentially toxic content before being dried and pounded into flour. This constituted a significant addition to existing labour demands and required adopting new techniques (Nweke *et al.* 2002; Haggblade & Zulu 2003). Cassava did however offer distinct advantages. The crop is resistant to drought, pests and disease, is not particularly labour intensive (apart from the planting and processing), provides food all year round (thus eliminating pre-harvest periods of hunger) and, most importantly, is relatively high yielding.[4] On the acidic Kalahari sands of Mwinilunga District, cassava can outperform sorghum, millet and maize yields.[5] As the eighteenth and nineteenth centuries progressed, cassava was therefore adopted as an increasingly prominent subsidiary crop, becoming the major staple in the area in the twentieth century.

The introduction and spread of cassava have previously been described in terms of an 'indigenous agricultural revolution'.[6] The crop's agronomic

---

[3] Interview with Mr Wombeki, Nyakaseya, May 2010.

[4] For an overview of the advantages and disadvantages of cassava cultivation, see von Oppen (1999: 51).

[5] The yields of different crops on sandy Kalahari soils were estimated as follows in 1962: maize: 3 bags per acre; millet: 4 bags per acre; cassava: 6.1 bags per acre. Source: National Archives of Zambia, Lusaka, hereinafter (NAZ) MAG1/20/4, Director of Agriculture Report, 17 May 1962.

[6] The label 'indigenous agricultural revolution' has been applied by von Oppen (1992) in connection with cassava but the phrase was first used by Richards (1985) who applied it more generally to other crops and agricultural changes.

characteristics, its effects on existing divisions of labour as well as its high-yielding potential brought about profound changes in the societies that chose to adopt the crop and Africanize its use (Vansina 1997). This Africanization involved ways of cultivating and preparing the root (constructing mounds, extracting the poison from the roots and making flour ready for consumption in the form of *nshima*) that were adapted to suit the particular Central African environment, existing methods of cultivation as well as local preferences regarding cuisine and taste. These methods of applying and using the crop differed significantly from how cassava was originally cultivated and prepared in its American area of origin, and could equally deviate from Portuguese customs and intentions behind the transmittal of the crop (*Ibid.*: 263-266).[7] The specific local environment thus transformed the way cassava was cultivated and consumed but cassava, in turn, equally transformed and altered existing agricultural and dietary practices in a continuous process of interaction, adaptation, appropriation and change.

Previous studies have argued that the crucial transformation that cassava brought about occurred mainly in the eighteenth and nineteenth centuries.[8] However it was only later that cassava shifted from being an important subsidiary crop to being the major staple crop in Mwinilunga District. Cassava's Africanization, which involved social, economic and political aspects in addition to the already-mentioned agro-ecological, labour and dietary issues, was a gradual process, the full consequences of which have only recently become evident. To understand this process, the political and economic context in which cassava cultivation was practised need to be considered.

[7] While the American method of preparing *farinha* meal from cassava never gained widespread popularity in the Central African region, methods of soaking cassava to remove its toxic content, of preparing a sauce from cassava leaves and of mound building were devised independently by Central African societies. These methods were unknown to American and Portuguese cultivators and could thus be labelled 'unintended' consequences of the introduction of cassava (Vansina 1999: 263-266).

[8] See Hoover (1978), Papstein (1978), Miller (1988), von Oppen (1992) and Vansina (1997). Von Oppen (1999: 59) argues that cassava went through yet another transformation in the twentieth century: 'Only then was the process which had begun in the nineteenth century of substituting cassava for millet (as opposed to the earlier supplementing of one by the other) completed'.

Photo 9.1 Cassava field with its rounded mounds [Photo: Iva Peša, 2010]

Engaging the state

Despite the advantages that cassava offered as a crop, colonial officials frequently criticized or even discouraged its cultivation and consumption. The alleged nutritional deficiencies, especially the acidity and lack of protein in the tubers, were regularly stressed (Jones 1959; Nweke 2002). The Lunda were described as being a 'weakly tribe' and 'physically inferior' due to their cassava-based diet.[9] Only when locust attacks or droughts ravaged the area did the colonial administration temporarily resort to promoting the crop: 'The cultivation of cassava is encouraged in the native areas throughout the Territory as a famine reserve crop . . . the crop came into its own during this difficult drought year'.[10] Nevertheless, attitudes towards cassava remained ambiguous at best, and just two years later negative valuations predominated once more: 'Methods of production of this crop are wasteful and it is preferable that it should be regarded as a subsistence rather than a cash crop.'[11] Instead of cassava, maize was promoted as a superior alternative. During the colonial period and espe-

9 (NAZ) KSE6/1/1, G.A. MacGregor, Balunda District Annual Report, 1909. 'On the whole the Balunda are a weakly tribe and physically much inferior to the natives of the Batoka District. Manioc [cassava] being the principle item of diet, and meat but sparingly indulged in . . . the Balunda are not good material from which to evolve good labourers. They are, upon the whole of very inferior physique, bone and stamina . . . The dietary of the Balunda is very limited apart from the manioc root, and this they prepare before eating, making a meal for porridge.'

10 (NAZ) Department of Agriculture, Annual Report, 1951.

11 (NAZ) Department of Agriculture, Annual Report, 1953.

cially after independence, maize was afforded prime importance by the government: 'As the staple food of the people of Zambia, maize is central in the Zambian agricultural economy'.[12] Various inputs, loans and subsidies were made available to support and encourage maize cultivation and highly favourable marketing opportunities for the crop were also arranged (Wood 1990; McCann 2005; Fourshey 2008). In contrast, cassava was afforded minimal attention by the government and was only marketed with difficulty through official channels. Officials regarded it as an inferior crop, though at times a necessary alternative to maize (Jones 1959).

This discrepancy can be explained, at least in part, by the fact that maize is a more easily administered and appropriated crop. Cassava, on the other hand, is relatively difficult to control, particularly from the perspective of the state that seeks to make resources accessible by creating 'a legible field of appropriation'. When taxation is collected, products that are easy to identify, monitor and enumerate (and can be readily assessed, collected and stored) are preferred (Scott 2010).[13] Seen in this light, cassava is relatively 'illegible' to the state. The crop flourishes under shifting cultivation in bush fields that may be a considerable distance from the village. In addition, it can prove difficult to ascertain the extent of cassava holdings as the best storage place is underground and harvesting takes place throughout the year according to the requirements of the cultivator. Early colonial administrators could, therefore, easily be deceived by cassava growers:

> (Headman Saloka) and his people appeared but denied that they had much food and brought just enough green cassava for one meal all round ... (Later) large cassava gardens were seen, the existence of which had been concealed by Saloka ... As we had no food I sent a party of twelve natives and two messengers to the gardens, to take enough green cassava for two days, as the gardens we had seen when leaving the village could amply provide this amount. While obtaining the food the carriers were attacked by Saloka and about fifteen of his people, armed with bows and arrows and guns.[14]

---

12 (NAZ) Review of the Operations of the Agricultural Marketing Committee, 30 June 1965.

13 Scott (2010: 73-74). Although the Asian context that Scott discusses is significantly different from the case of Mwinilunga, especially because regulations related to taxation and resource extraction in Mwinilunga were never strictly enforced by the state and taxes were paid in money rather than as a share of harvested crops, Scott's insights highlight the ways in which cassava served as a tool to engage (or in some cases avoid engagement) with the state.

14 (NAZ) KSE6/6/1, Captain Stennett, Balunda District Report, 12 January 1910.

In this case though the administrative party might have confiscated both mature and immature cassava, which makes the violent reactions of the cultivators more understandable because the uprooting of immature cassava could potentially jeopardize future food supplies. It is equally evident that the untrained eye can easily fail to judge the extent of crop holdings. Since the size and content of cassava gardens were difficult to ascertain and even more difficult to appropriate, negative value judgements were attached to the crop, especially by government officials.

The 'slow' nature of cassava proved to be an equal hindrance.[15] The varieties grown in Mwinilunga District take between 18 and 36 months to mature but can be left in the ground for up to 48 months before harvesting. From an administrative point of view, annual crops, such as millet or maize, are easier to assess and control and are therefore preferable to cassava that takes longer to mature. In addition, the processing of cassava roots is a process that lasts several days. While grain crops can be kept in storage bins and processed into flour within several hours upon request, cassava roots require soaking and drying and only then are they ready to be pounded into flour. This generated irritation among touring colonial officers:

> The absence of grain (sorghum and millet) in the District makes the transit of a large party a matter of difficulty. No emergency supplies are in hand at the villages, and the preparation of flour from the manioc (cassava) root is a long process extending over several days.[16]

Delays ensuing from the processing of cassava could be distinctly advantageous to the village population though. The ingenious excuses devised to delay pre-colonial trading caravans have been well documented. Traders were charged with imaginary offences, men were inebriated or seduced, and all kinds of plots were devised so that the caravan remained in a village for a few additional days, which would give the population the opportunity to trade and purchase certain goods (Pritchett 2001; Rockel 2006). In a similar vein in the colonial period, hold-ups caused by cassava processing gave the village population a chance to retain officials in their surroundings for some time. While women were pounding cassava, the headman could discuss colonial laws and policies, enquire about job opportunities, sell food and create a generally favourable image of the

15 The concept of 'slowness' is taken from Peveri (2011) who described enset cultivation in Ethiopia.

16 (NAZ) BS2/199 IN2/1/3, G.A. MacGregor, Balunda Sub-District Monthly Report, September 1908.

village for the visiting colonial officer. Cassava was sometimes deployed as a tool to negotiate (favourable) involvement with the colonial state.

The role that cassava could play in state engagement or evasion was worked out in the early part of the twentieth century when the move was made to establish colonial rule in Mwinilunga District. Several shows of violence, especially under the administration of G.A. MacGregor in 1908-1909 and as a result of the introduction of taxation in 1913, accompanied the build-up of an administrative presence in the area. Gardens were raided to obtain food, huts were burnt and people were imprisoned or even shot (Macpherson 1981; Pritchett 2001; Peša 2009). These sporadic displays of violence should not be overemphasized however as the colonial administration was numerically weak and their hold on the area remained tenuous. Nevertheless, the administrative presence did encourage some individuals to seek their fortunes elsewhere, usually temporarily but sometimes permanently. Some fled into the bush for days at a time and returned as soon as the officials had moved on while others joined relatives in Angola or Congo for months or even years.[17] According to some accounts, two-thirds of the population fled following the introduction of taxation. The consequences of this exodus were described in dramatic terms:

> The running last May has caused consequent starvation and the people are getting restless because of this. For these people were not content merely to run and leave their gardens to be eaten by the wild pigs in the bush – but they must also ply the hoe in the hand too and root up even their young cassava, so that they could leave with the feeling of finality and of burnt ships behind them. Now that most of them are back they are starving, and those who did not run are suffering.[18]

The severity of events is probably overstated by this account considering that two years later the majority of villages had resumed settlement in the area and abundant harvests as well as sales of food were reported.[19] Nevertheless, the role of cassava in the flight and subsequent readjustment to settlement deserves closer attention.

---

17 See also Peša (2009: Chapter 3).

18 Bodleian Library of Commonwealth and African Studies, Rhodes House, Oxford, UK, hereinafter (BOD) Mss Afr. S779, Theodore Williams's correspondence, 16 February 1914.

19 For example, (NAZ) KSE6/1/1, Captain Stennett, Balunda District Tour Report, August 1909 (NAZ) KSE4/1, Mwinilunga District Notebook; (BOD) MSS Afr. S776, Theodore Williams's diary, 23 June 1914.

Since the crop is not tied to a particular harvesting season, extra supplies could be dug up when embarking on a journey to provide food along the way as well as to ensure a warm welcome from friends and/or relatives visited. If individuals decided to return to their villages within several months or even a year, cassava fields would probably be found intact and ready for harvesting. Grain crops, on the other hand, had to be harvested at once at specific times of the year. A travelling individual, if grain was available in the storage shed, had the choice of carrying all the grain available on the journey or of leaving the shed to the mercy of raiding officers, wild animals or greedy neighbours. Cassava was far more difficult to raid as it first had to be dug up, which required a concerted effort. Once the grain cultivator returned to his village, supplies of food would be depleted. New crops would have to be planted and only when these matured would food be available again. In this respect, cassava facilitated mobility and evasion of the administration (Scott 2010: 178-181). Preventing hunger and facilitating flight, cassava allowed a relatively independent, flexible and, at times, defiant stance towards the newly established colonial state.

Taken together, these examples suggest that cassava could be deployed as a tool to engage the state on favourable terms and one that offered individuals space for manoeuvre. By withholding food or fleeing, individuals achieved negotiating power *vis-à-vis* colonial officers, much to the exasperation of the latter:

> Little or no food is at present being brought to the Boma for sale ... (This) is the more annoying when it is remembered how large are the gardens and plentiful the supply of cassava ... the present position is altogether attributable to the perverseness of a people endowed with so peculiar and erratic a temperament.[20]

It was also stated that 'There may be a determination not to come under our rule ... to escape from obligations which they fear will be demanded from them'.[21] Local resistance could, thus, counteract administrative goals and proved remarkably powerful. In the case of Mwinilunga District, cassava was used by individuals as a tool to assert autonomy and was therefore associated with 'inferiority' in the eyes of the administration. Compared to the easily controllable and orderly grain crops, cassava seemed a 'rebellious' crop.

20 (NAZ) KSE6/5/1, J.M. Pound, Balunda District Monthly Report, June 1909.

21 (NAZ) BSA2 A2/1/4 Loc. 3981, Acting Administrator North-West Rhodesia to G.A. MacGregor, 20 December 1910.

Photo 9.2 Pounding and preparing cassava flour
[Photo: Sakeji School pictures, courtesy of Dennis Brubacher (n.d. but from the 1950s)]

Stepping into the market

Cassava similarly enabled a relative autonomy when compared to other crops with respect to marketing. This autonomy was often negatively linked to producing for self-subsistence, or rather a lack of market initiative. In the early twentieth century, it was described as follows by colonial officials: 'They have no conception of providing for beyond their own requirements. The calls of travellers for food upset their calculations, and food is reluctantly provided in small quantities at exorbitant charges, or rather valuations, or refused altogether.'[22] Government reports from the 1970s seemed to express similar grievances: 'Most of the people are still subsistence farmers, growing enough only for their consumption requirements, and only selling a little which enables them to purchase basic household utensils.'[23] In spite of this supposed apathy towards market production, up to 600 tons of cassava flour were marketed per year in the 1950s.[24]

---

22 (NAZ) KSE6/1/1, G.A. MacGregor, Balunda District Annual Report, 1908-1909.

23 (NAZ) MRD1/8/27 Loc.4272, North-Western Province Development Committee, 20 March 1970.

24 (NAZ) NWP1/2/68 Loc.4911, North-Western Province Annual Report for Agriculture, 1955. Considering that there were approximately 35,000 inhabitants in the district at the time, these sales figures are relatively high. Throughout the early colonial period,

Cassava was like a store of wealth in the ground. When travelling through the area in the 1850s, David Livingstone remarked how:

> Very little exertion is required to procure the staff of life, which in these parts is the manioc (cassava) . . . (It provides) a supply of food for years. The climate is so good, they are either planting or reaping the whole year round. (Schapera 1961: 261-262)

A certain reserve of cassava was always held back for visiting relatives, refugees or for marketing. Cassava roots are only harvested when required and part of the crop is never harvested at all (Pritchett 2007: 195).[25] It therefore provides an emergency back-up in times of need or, alternatively, the crop can provide an entry into the market. When marketing opportunities for cassava prove favourable, individuals can produce a small surplus from what is already available in the ground without too much difficulty and if markets remain profitable, cultivators can expand their fields in subsequent harvests and consume any surplus within the village if the market collapses. Alternatively the cassava can be stored in the ground until prices rise once more. Cassava is a highly flexible crop regarding market and price fluctuations and is 'well suited to both subsistence and commercial farming, on either a small or a large scale' (Berry 1993: 187).

Maize, another introduction from the New World, was fervently promoted by government policies and afforded preferential marketing. Government-controlled marketing boards, especially following independence, provided secure and profitable outlets for the crop. The fact that these gained a near monopoly on formal rural marketing yet generally refrained from buying cassava may have persuaded cultivators in Mwinilunga District to shift from growing cassava to growing maize.[26] This move, however, proved to be far from automatic, straightforward or universal. Complaints concerning the lack of maize cultivation in the area were frequently voiced: 'any (maize) surplus marketed was merely a fortuitous

sales of cassava flour remained limited to administrative and missionary requirements but in the 1940s and 1950s larger amounts of cassava flour were sold to traders who transported it to markets in Congo or urban areas along the Zambian line-of-rail, where demand soared as a result of developments (and increased employment) at the copper mines.

25 Pritchett (2007: 195) suggests 'that at least one-third of the available cassava is never dug up'.

26 Maize rather than cassava was adopted on a large scale in many other areas across Zambia. For a comparative case study on maize adoption in Tanzania, see Fourshey (2008).

surplus from subsistence cultivators and did not constitute an appreciable amount'.[27] Although 180 tons of maize was marketed in Mwinilunga in the 1969/1970 season, this figure compared unfavourably with other districts, and the country as a whole.[28]

Instead of attributing the drop in production to a lack of initiative on the part of rural producers, a plausible explanation may be found in the preference for cassava and the relative autarky provided by the crop. Relying solely on maize cultivation would have been a gamble as the crop does not grow particularly well in the local climate and the soils of Mwinilunga, which means that there is a constant threat of harvest failure. In addition, expensive inputs such as fertilizers, pesticides and improved seeds are imperative to obtaining yields that can compete on the market. These inputs result in individuals being dependent on both the market and the state. Government-controlled marketing boards provided inputs on a loan basis and after selling one's maize the individual first had to think of repaying his debt:

> If a farmer sells two bags to (the) Agriculture Rural Marketing Board (ARMB) then the very farmer should be made to repay his loans out of what he has just got from those two bags . . . but how can a farmer live since all his living is taken away from him/her? Now we have a lot of groaning the farmers are murmuring perhaps in future they will stop doing farming because all the money they are getting from their produce is being taken away from them without leaving them some Ngwees to enable them buy clothing, soap, salt, and paying their friends who helped them to do the job (sic).[29]

Although maize might have been a commercially viable option and was increasingly adopted as a cash crop in the 1960s and 1970s, it left individuals vulnerable to the whims of the climate, state subsidies and the marketing boards.

Whilst cassava did not offer equally encouraging marketing opportunities, the crop did provide food security throughout the year. Villagers proudly remarked that 'with cassava we never go hungry'[30] as the crop eliminates hunger in the pre-harvest period when grain supplies become depleted. However, cassava could equally function as a cash crop if markets were available. For example, large amounts of cassava from

27 (NAZ) Annual Report, Ministry of Agriculture, 1964.
28 (NAZ) LGH5/5/12, North-Western Province Marketing of Produce, 23 July 1970.
29 United National Independence Party archives, Lusaka, Zambia, hereinafter (UNIPA) UNIP5/3/1/48, J. Chikotola to Administrative Secretary, 1 October 1973.
30 Interview with Justin Kambidimba, Ntambu, September 2010.

Mwinilunga District were sold to urban areas in the 1950s, to Angolan refugees fleeing the civil war and, in 1982, marketing boards resorted to buying cassava due to failing maize harvests (Pritchett 2007: 195, 202). Since cassava functioned as both a food and a cash crop, reserves were already at hand and could be used to step into the market relatively easily. At the peak of cassava sales in 1949, it was stated that 'the most remarkable feature of the agriculture of the district is the tremendous surplus of cassava meal, which becomes greater year by year'.[31] No expensive inputs were needed to expand cassava output so cultivators were independent of outside agencies for loans and fertilizer. The main requirement was labour and although it was scarce, once marketing opportunities arose, a household unit retained the choice of deploying some of its labour force into cassava production, while the remainder could focus on alternatives such as labour migration or hunting.[32]

The fact that cassava production for the market could exist alongside labour migration ratios of 50% in the 1950s is clear evidence that cassava sales did not necessarily have to be a strain on the limited labour resources.[33] Instead, and judging by the large quantities sold, it was distinctly attractive and potentially lucrative to produce cassava. Far from lacking market logic, cassava cultivators initiated sales of their harvests but equally opted to retain a dependable supply of food to hand rather than submitting completely to the vagaries of fluctuating markets.[34] Unlike

31 (NAZ) SEC2/156, R.C. Dening, Mwinilunga District Annual Report on African Affairs, 1949.

32 Moore & Vaughan (1994: 155-156) provide the following assessment for the case of Zambia's Northern Province: 'Some of the increase in agricultural production was due to communities putting large acreages under cassava. There are many different types of cassava with varying maturation times, but one common variety grown in the province was harvested only after three years. Such a crop is clearly less labour intensive than is the finger millet grown on citemene fields. However, it must be remembered that cassava was grown on mounds, and that the labor required for mounding was very substantial, particularly if acreages were large and continuously being extended. It is not enough, therefore, to claim that cassava solved the problem of labor shortages, although it is true that its production did not necessarily require male labor.' Some authors have argued that cassava is a labour-saving crop and thus facilitated male labour migration as women could clear and work garden plots by themselves. The processing of cassava into flour is, however, relatively labour intensive (Berry 1993: 187-189).

33 (NAZ) NWP1/2/102 Loc.4919, E.L. Button, North-Western Province Annual Report, 1960.

34 For a similar argument acknowledging the importance of flexibility and independence in crop production, see Hyden (1980: 29, 194): 'Much of the power of the small peasants in Africa stems from their control of the means of subsistence. The production of the basic necessities is still controlled by peasants who are difficult to get at, not only because of their numbers but also because they are capable of securing their own

maize, cassava enabled individuals to participate in market production independently when this proved profitable and/or desirable. Although cassava might not have provided stable or extensive markets, the crop could always be relied on as a staple food if markets slumped. Cassava offered stability but could be expanded at will and still relied upon in times of hardship. Although maize was increasingly adopted as a cash crop after independence and the acreage planted to cassava decreased somewhat, the cultivation of cassava was never abandoned. The crop withstood locust epidemics, marauding wild animals and drought, and remained the area's food of choice throughout the twentieth century.

Transformation, preferences and comparison

The transformations cassava brought about in the eighteenth and nineteenth centuries influenced society profoundly:

> The most important increases in African productivity during the era of import growth, and the major technological transformations in the African

Photo 9.3 Girl with cassava sifting and storage baskets
[Photo: (NAZ) SEC2/964, Mwinilunga District tour report accompanying photographs (n.d. but from the 1950s)]

subsistence and reproduction without the assistance of other social classes ... peasant farmers producing export crops continue to grow most of the food they consume.' Hyden makes a case for an 'uncaptured peasantry' that is relatively autonomous of the state and market domination.

> political economy, took the form of incorporating new American food crops, principally manioc (cassava) and maize, into the agricultural systems of the area crossed by the slaving trails. These exceptions eloquently demonstrated the general rule that otherwise limited 'development' in the technological sense. Manioc and maize were 'appropriate' technology in their negligible acquisition cost to farmers, required virtually no long-term capital investment to implement, and fit well into the existing strategy of raising dependents at home, within the autoconsuming communities of western central Africa, since they enabled landowners to feed more people from the same area than the old crops would have supported. (Miller 1988: 103)

Adopting cassava was an example of technological innovation that touched the core of agricultural production and, in the long run, even the social, economic and political organization. The previous sections have argued, however, that cassava adoption on a large scale set other innovations and transformations in motion in the twentieth century. Only then was cassava adopted almost universally across Mwinilunga District and adapted to existing agricultural patterns of production, transforming aspects of societal organization in the process as it became fully appropriated. The impact of these transformations and the interwoven relationship between agriculture, the economy, politics and aspects of identity is reflected in the Lunda saying '*makamba mwanta wetu*' (cassava is our chief). However, if cassava was indeed such an attractive and flexible crop, why did other Zambian and Central African producers refrain from adopting it on a similar scale? And from a comparative perspective, why did maize gain such a strong foothold in other areas whilst cassava remained a 'famine crop' at best?[35]

Under colonialism as well as after independence, 'an image of maize as a model of modernity' (Fourshey 2008: 246) was created and the high-yielding hybrid varieties were acclaimed as being 'magic' or a 'technological wonder' that would 'feed the nation' and bring about general prosperity (Pottier 1988: 20). In the case of Tanzania, British colonialism created an image of maize as the quintessential cash crop, a tool with which to step into the capitalist economy and that would undoubtedly bring progress and modernity:

---

[35] For comparative examples in the Zambian context, see Pottier (1988), Moore & Vaughan (1994) and Vickery (1984/1985). The adoption of cassava or maize varied per region: producers in the Southern Province, for example, adopted maize and largely rejected cassava.

> The state intended to persuade farmers to create more densely populated settlements, to increase production yields by growing a single crop on larger parcels, to incorporate more wage labor in farming methods, and to shift to crops that had greater demand in the world markets ... The state used maize as a cash crop to tighten ties to currency ... increase tax revenue, and increase chances that production would reach the 'appreciably' higher levels of surplus and prosperity that the state expected farmers to achieve. In order to restructure the economy, the state needed to convince people to consume a new crop like maize, which the state could control. (Fourshey 2008: 235, 243)

Crop production was evidently not merely connected to individual preference or ecological suitability but hinged on issues of state control, marketing and capitalism. Maize may have failed to guarantee general food security – as harvests tended to fluctuate dramatically and were dependent on expensive inputs – but could be marketed and thus provided cultivators with prospects of prosperity (Pottier 1988; Fourshey 2008). Maize production involved market interaction and state control, both of which were potentially profitable and beneficial but could equally entail risk. While these risks seemed acceptable to some, notably to producers close to urban areas with easy market access, to others they did not.

Although cassava was not a particularly lucrative crop and as a consequence was unattractive for market-oriented producers, it could minimize any associated risks by providing food security. As argued above, producers in Mwinilunga District adopted cassava because it gave them flexibility and independence regarding markets as well as towards the state. And though cassava was generally regarded as a 'backward' crop that was only suitable for household or subsistence production, it afforded producers distinct agency. Mwinilunga District enjoyed poor access to major markets and was not connected to the Copperbelt by a good road network. This raised transport costs considerably and made the marketing of low-value high-weight goods, such as cassava, generally unfeasible. Although the marketing of maize, a crop which fetches higher prices than cassava on the market, continues to be attempted in the area, success is hampered by transport costs, access to markets and a lack of inputs. Despite being a symbol of modernity in the eyes of the state and profitable in terms of marketing, maize was not unequivocally attractive to producers in Mwinilunga District. To them, it remained an alien symbol of dependency on the state and fluctuating markets. Even in the late 1950s the crop was only hesitantly being accepted: 'People are not at present interested in eating maize themselves unless cassava is scarce, they claim indeed it makes them sick'.[36] The flexibility and autonomy that cassava facilitated proved to be a distinct asset and enticed producers to choose cassava instead of maize.

Cassava is a flexible crop in more than one sense. Numerous varieties are grown interchangeably or even in the same field, with varieties differing in taste, resistance to disease, length of maturation, leaf growth, yield and level of toxicity in their roots (Jones 1959; Nweke 2002). A survey from the 1940s estimated that within one village, 28 varieties of cassava were being grown simultaneously.[37] This was the result of travellers or migrant labourers returning from urban areas or Congo and bringing back a cassava stem to plant in their home village. The variety would subsequently be named after the traveller who first planted it and led to varieties known as 'Chisamukila', 'Kambongi' or even 'Mary'.[38] A new variety would be tested by intercropping it in a field with other varieties and crops, and it was only retained if it demonstrated favourable characteristics.[39] The proliferation of varieties of cassava minimized the risk of disease or crop failure because if one variety was struck by disease, another might be unaffected. Although today the high-yielding and fast-maturing Bangweulu variety is gaining popularity throughout the area, other varieties continue to be grown too.[40] Cassava has never been monocropped in fields but a cropping pattern has emerged whereby sorghum or millet is planted in the first year when gardens are being cleared. In the second and subsequent years, cassava mounds are constructed and intercropped with vegetables, legumes and other crops (Trapnell & Clothier 1957: 5-6). A single cassava field can thus provide a fairly balanced and varied diet in and of itself, as it will regularly produce beans, groundnuts and even maize. Allowing a varied diet, spreading risk and providing general food security, cassava has remained a popular and preferred source of food across Mwinilunga District.

36 (NAZ) SEC2/967, W.D. Grant, Mwinilunga District Tour Report, No. 5, 1959.

37 (NAZ) SEC2/955, R.C. Dening, Mwinilunga District Tour Report, November 1947.

38 Interview with Mr Kapaipi, Ikelenge, March 2010. See also (BOD) Richard Cranmer Dening's uncatalogued papers (12 boxes).

39 It could be postulated that the introduction of cassava into the area was encouraged by a mixture of mobility, experimentation and agricultural innovation similar to that characterizing the subsequent introduction of different varieties.

40 Interview with Ambrose Musanda, Mwinilunga, October 2010.

Photo 9.4 Young boys consuming a cassava-based meal
[Photo: Sakeji School pictures, courtesy of Dennis Brubacher (n.d. but from the 1950s)]

A marker of identity

The multiple advantages of cassava have resulted in the inhabitants of Mwinilunga today proudly referring to themselves as 'cassava eaters'.[41] This sense of pride is countered by remarks from urban dwellers who pejoratively note the 'bubble gum' consistency of cassava *nshima* and denounce cassava consumers as being 'backward'.[42] Nonetheless, cassava has been incorporated into the heart of Lunda identity.[43] It is appreciatively described as an all-in-one crop as the tubers provide food, the stems are used as planting material, the stalks function as firewood and the leaves form a nutritious and tasty vegetable dish.[44] Cassava is so highly valued in the area that it has even been incorporated into ceremonies and rituals. Cassava flour is now a substitute for the white clay powder (*mpemba*) formerly used in rituals, symbolizing 'whiteness' or 'purity' (*kutooka*, to be white, clean, or pure). More specifically:

---

41 Interview with Solomon Kanswata, Mwinilunga, March 2010.

42 Such remarks were common in Lusaka or on the Copperbelt when talking about people who consumed cassava *nshima* on a daily basis.

43 For a useful analysis of cassava production, its cultural meaning and Aluund identity, see de Boeck (1994: 257-282).

44 Interview with John Kamuhuza, Ikelenge, March & April 2010.

> Whiteness (*wutooka*) in Ndembu ritual has many connotations. These include: strength, health, good luck, ritual purity, authority, good will between the ancestors and the living, the clear and known as opposed to the obscure and unknown, life, power, breast milk, seminal fluid, the whiteness of cassava meal and roots. (Turner 1962: 49)

Cassava is commonly regarded as a 'female crop' because women execute the main tasks of cultivating and processing, and it also figures prominently in female initiation and healing ceremonies (Turner 1957; Pritchett 2001). Although cassava may have increased the workload of female cultivators, it has equally provided them with opportunities to partake in market production and reduced their dependency on male labour for the felling of trees and the preparation of fields.[45] While new fields have to be cleared yearly for grain crops, cassava fields can be cultivated for up to twenty years without any fertilization being required. The relative autonomy and agency afforded by cassava cultivation to women, but equally to entire villages and communities, has made cassava a distinct asset and an integral marker of identity for the people of Mwinilunga.

Conclusion

Cassava, initially a foreign introduction to the area, has proved itself to be a uniquely versatile crop. While it was used to aid Lunda state formation in the pre-colonial period, it was adopted as a tool to defy onerous demands from colonial officers during the colonial period. The crop has provided individuals with manoeuvring space and allowed negotiation on favourable terms with the state and markets. Cassava has become a symbol of a separate identity that is labelled 'backward' by some, but is seen as a source of pride by others. The crop has infiltrated the very core of Lunda society and has been adopted in numerous rituals and deployed as a symbol of group coherence. Cassava was thoroughly Africanized and has served to unleash an enormous potential for production in the area, to the point that Mwinilunga has become a

[45] (NAZ) SEC2/154, Western Province Annual Report, 1947. 'Where cassava is the staple crop and agriculture is largely stabilised, the presence of the man is unnecessary for the preparation of the gardens.' Cassava cultivation might be viewed as a double-edged sword. The crop freed up male labour from agricultural production and created a degree of independence for women. The sale of surplus cassava has offered women an income-generating opportunity but has also greatly increased female workloads in mound building and processing. Some have even argued that cassava has 'enslaved' women. See Spring & Hansen (1979: 1-22). Cf. Pritchett (2001: 50-51), von Oppen (1992: 283-285) and Vansina (1997: 275-276).

'land of plenty', boasting large cassava gardens for home consumption or external sales. In this sense, cassava has enabled the inhabitants of Mwinilunga to establish and secure their own distinct existence. Today, cassava remains the area's favourite food as people proudly proclaim: 'Cassava is our staple food, cassava is our chief!'[46]

References

Berry, S.S. (1993) *No Condition is Permanent: The Social Dynamics of Agrarian Change in Sub-Saharan Africa*, Madison: University of Wisconsin Press.

Crosby, A.W. (1972) *The Columbian Exchange: Biological and Cultural Consequences of 1492*, Westport: Greenwood Press.

de Boeck, F. (1994) '"When Hunger Goes Around the Land": Hunger and Food among the Aluund of Zaire', *Man* 29(2): 257-282.

Fourshey, C.C. (2008) '"The Remedy for Hunger Is Bending the Back": Maize and British Agricultural Policy in Southwestern Tanzania 1920-1960', *The International Journal of African Historical Studies* 41(2): 223-261.

Haggblade, S. & B. Zulu (2003) 'The Recent Cassava Surge in Zambia and Malawi', Paper presented at the Successes in African Agriculture conference, Pretoria, South Africa, 1-3 December.

Hoover, J.J. (1978) 'The Seduction of Ruwej: Reconstructing Ruund History (The Nuclear Lunda; Zaire, Angola, Zambia)', Unpublished PhD Thesis, Yale University.

Hyden, G. (1980) *Beyond Ujamaa in Tanzania: Underdevelopment and an Uncaptured Peasantry*, Berkeley/Los Angeles: University of California Press.

Jones, W.O. (1959) *Manioc in Africa*, California: Stanford University Press.

Macpherson, F. (1981) *Anatomy of a Conquest: The British Occupation of Zambia, 1884-1924*, Harlow: Longman.

McCann, J.C. (2005) *Maize and Grace: Africa's Encounter with a New World Crop, 1500-2000*, Cambridge: Harvard University Press.

Miller, J.C. (1988) *Way of Death: Merchant Capitalism and the Angolan Slave Trade 1730-1830*, Madison: University of Wisconsin Press.

Nweke, F.I., D.S.C. Spencer & J.K. Lynam (2002) *The Cassava Transformation: Africa's Best-Kept Secret*, East Lansing: Michigan State University Press.

Papstein, R.J. (1978) 'The Upper Zambezi: A History of the Luvale People, 1000-1900', Unpublished PhD Thesis, University of California.

Peša, I. (2009) 'Cinderella's Cassava: A Historical Study of Agricultural Adaptation in Mwinilunga District from Pre-colonial Times to Independence', Unpublished MPhil Thesis, Leiden University.

Peveri, V. (2011) 'Enset, the Tree of the Poor: Nutrition and Identity in Hadiya Zone (South-central Ethiopia)', Paper presented at the ECAS Conference, Uppsala, 15-18 June.

Pottier, J. (1988) *Migrants No More: Settlement and Survival in Mambwe Villages, Zambia*, Manchester: Manchester University Press.

Pritchett, J.A. (2001) *The Lunda-Ndembu: Style, Change, and Social Transformation in South Central Africa*, Madison: University of Wisconsin Press.

—— (2007) *Friends for Life, Friends for Death: Cohorts and Consciousness among the Lunda-Ndembu*, Charlottesville: University of Virginia Press.

Richards, P. (1985) *Indigenous Agricultural Revolution: Ecology and Food Production in West Africa*, London: Hutchinson.

---

46 Interview with Julian Chiyezhi, Mwinilunga, May 2010.

Rockel, S.J. (2006) *Carriers of Culture: Labor on the Road in Nineteenth-Century East Africa*, Portsmouth: Heinemann.
Schapera, I., ed. (1961) *Livingstone's Missionary Correspondence 1841-1856*, London: Chatto & Windus.
Schecter, R.E. (1976) 'History and Historiography on a Frontier of Lunda Expansion: The Origins and Early Development of the Kanongesha', Unpublished PhD Thesis, University of Wisconsin-Madison.
Scott, J.C. (2010) *The Art of Not Being Governed: An Anarchist History of Upland Southeast Asia*, New Haven/London: Yale University Press.
Spring, A. & A. Hansen (1979) *Women's Agricultural Work in Rural Zambia: From Valuation to Subordination*, Waltham: African Studies Association.
Springer, H.E. (1909) *Snap Shots from Sunny Africa*, New York: The Methodist Book Concern.
Trapnell, C.G. & J.N. Clothier (1957) *The Soils, Vegetation and Agricultural Systems of Northwestern Rhodesia: Report of the Ecological Survey*, Lusaka: Government Printer, 2nd edition.
Turner, V.W. (1957) *Schism and Continuity in an African Society: A Study of Ndembu Village Life*, Manchester: Manchester University Press.
—— (1962) 'Themes in the Symbolism of Ndembu Hunting Ritual', *Anthropological Quarterly* 35(2): 37-57.
Vansina, J. (1997) 'Histoire du manioc en Afrique Centrale avant 1850', *Paideuma* 43: 255-279.
Vickery, K.P. (1984/1985) 'Saving Settlers: Maize Control in Northern Rhodesia', *Journal of Southern African Studies* 11(2): 212-234.
von Oppen, A. (1992) '"Endogene Agrarrevolution" im vorkolonialen Afrika? Eine Fallstudie', *Paideuma* (38): 269-296.
—— (1999) 'Cassava, "The Lazy Man's Food"? Indigenous Agricultural Innovation and Dietary Change in Northwestern Zambia (ca. 1650-1970)'. In: C. Lentz, ed., *Changing Food Habits: Case Studies from Africa, South America and Europe*, New York: Routledge, pp. 43-72.
Wood, A.P., ed. (1990) *The Dynamics of Agricultural Policy and Reform in Zambia*, Ames: Iowa State University Press.

10

The social cocktail: Weddings and the innovative mixing of competences in Botswana[1]

Rijk van Dijk

Weddings in Botswana are increasingly being marked by emerging complexities for a younger generation in the way these events draw upon different social, cultural, economic, legal and moral sources and resources. Since this complexity represents a new mixture of sources that are local and global, traditional and modern, and part of a civil system as well as part of the rising consumerist style, the people involved require a new set of skills and competences to ensure their wedding is a success. With weddings emerging as a new social cocktail, it is difficult to identify whether the elements are traditional by origin, have been adapted from elsewhere or are entirely new innovations. This contribution argues that much of the transformative innovation is located in the rise of these skills and competences. It identifies four interlocking domains that are part of the organization and styling of weddings for which an innovation in the personalized skills and competences of couples has become crucial. Although rituals are always adaptive to changing social environments, the rise of these new skills and competences is not merely adaptive but changes social, gender, economic and age-related structures as is manifest in these weddings.

[1] I am indebted to colleagues at the University of Botswana, in particular Musa Dube, Godisang Mookodi & Doreen Setume as well as my research assistants John Hamathi & Bernard Rantuana, for their support and interest in my work in Molepolole. I am also grateful to the convenors of the workshop on 'Innovation in Africa' and to Astrid Bochow, Marloes Janson, Julia Pauli & Hansjoerg Dilger for their comments on an earlier version of this chapter.

Introduction

In bookshops in some of the urban centres of Botswana, such as the capital Gaborone but also in smaller towns such as Molepolole and Mochudi, there are increasing numbers of glossies portraying the splendour and grandiose style of so-called white weddings. These have become very popular and are innovatively instilling Western and global upper-middle-class lifestyles and appetites within local settings. While these weddings seem to be responding to consumerist ideas of the arrangement of relationships in a modern African society (for Nigeria, see Bastian 2001), they emulate the emergent interests of a younger generation and are simultaneously responding to traditional cultural requirements, the strictures of the country's civic system and Christian moral notions. In the peri-urban agro-town of Molepolole (an hour's drive from Gaborone), what can be seen emerging in these costly, lavishly-styled weddings is a highly complex 'cocktail' of arrangements, appetites, styles, formal requirements and ritual procedures that are difficult to disentangle. This complexity makes it analytically challenging to distinguish and classify elements of local culture according to 'tradition', 'innovation' or 'adaptation'. Where does one category begin and the other end?

In being a 'social cocktail', these weddings challenge social science in perceiving or maintaining such categories of analysis. Yet they also challenge the young couples themselves as these weddings signal a mix of constraints and the competences they need to acquire in dealing with the entire process. None of these styles, appetites or traditional requirements come uncontested or effortlessly. This is partly because in a locality such as Molepolole, weddings are part of what Appadurai (1996) has called a production of locality in which a precise 'mixture' of what are considered 'local customs' and engagements with a wider world of style, aesthetics but also citizenship and class aspirations are being pursued. Such a production of locality – in the context of this contribution, the production of local ritual ceremonies that constitute a marriage – requires detailed knowledge of how and what to mix. The innovative capacity these groups espouse, it is argued here, must be analyzed in terms of how new competences emerge in the negotiation of a tantalizing social cocktail.

Exploring issues of adaptation and innovation should also focus not only on that which is being produced (i.e. the innovative artefact, object or technique) but also on its reception, contestation and (internal) contradictions (Ferguson 1999). In the case of these weddings, the competences that are required in order to deal with all of these dimensions are relatively new. They include, for example, being able to live up to the expectations of one's elders and the guardians of cultural traditions, in addition to registering one's marriage under Roman Dutch law, and marrying according to class-based peer pressure.

While the development of these weddings and their increasing middle-class styling might be termed 'adaptive', the rise in the level of competences required and the contestations that are embedded can be termed 'innovative'. There is a qualitative difference between adaptation and innovation that may not only imply an aspect of time and social change but also an element of the way in which a wider social environment is affected, involved or changed in the process. It is about whether innovation arrives because of adaptation to a (changing) (social) environment or occurs by changing the (social) environment itself. This indicates a qualitative difference in the nature and direction of innovation. In the case of these weddings, this chapter argues that a qualitative change in the competences that are required for dealing with the intricacies of their complexity is the innovative aspect of these ceremonies. While rituals are, arguably, always adaptive to changing social environments, the rise of new skills and competences are having an effect on changing an environment of social, gender, economic and age-related structures.

The social cocktail of requirements: Weddings in Botswana

As is the case in many places in Africa (Mann 1985; Cornwall 2002; Masquelier 2004; Johnson-Hanks 2007; Pauli 2009; Constable 2009), weddings in Botswana are becoming hugely expensive affairs and the costs are now being borne by the couples themselves. In Molepolole for example, meeting the expectations of their families as well those related to their new middle-class positions may lead couples to have weddings that cost anything up to BP 200,000 or more (i.e. three times their annual income for some). It is important to note that for many in Molepolole (the couples as well as their families), a dramatic change has occurred in the way weddings are being celebrated and the marital relations that are formed. As Schapera (1940, 1950) saw in the 1930s and 1940s in nearby Mochudi, weddings and marital relations existed in a variety of forms. This variety was partly caused by massive labour migration by Bechuanaland men to the South African mines, particularly in Johannesburg, that reshaped not only the marital economy but also the long-distance aspects of a marital relationship (Brown 1983; Townsend 1997). Making money in the mines meant that men, on their return to their homesteads, were in a position to start spending more on things like wedding outfits, the *lobola/bogadi* (bride price) or a neo-local residence for the couple. Over the years, another major change was the gradual disappearance of polygamous marriages as well as arranged marriages (Solway 1990), partly under the impact of Christianity and colonial rule and partly because of education and the subsequent changes in the socio-economic positions of men *and* women (Suggs 2002; Good 2008).

The well-known dynamics of the changing nature of marital arrangements and economies are important when stressing that marriage has always been dynamic and subject to change, and that marital arrangements have a specific constellation or coming together of various traditions, interests, social developments and styles. Perhaps as no other social institution, marriage in Botswana can be seen as being constantly subjected to innovation and as adapting to changing socio-economic, cultural and generational conditions. In the Molepolole context, where I have studied the wedding arrangements of 25 couples and their families, these weddings can be seen as conflating four different fields of social interaction, power-relations and moralities.

The first that can be disentangled are the wedding traditions and their transformations over time in view of what people today identify as belonging to 'our culture'; i.e. a set of ideas concerning traditions and customary practices (*ngwao*). While ideas of how marriage should come close to and represent notions of culture and tradition are replete with nostalgia, this nostalgia is at the same time not without power. It is usually a power much vested in the position of the elderly, broadly categorized as 'parents', that is at play. These crucial elements for the proper execution of weddings involve such things as the payment of bride wealth (*lobola/bogadi*), the involvement of maternal uncles who are in charge of the wedding negotiations at the *patlo* (the formal request and prestation[2] of important gifts to the bride's family), the involvement of traditional healers (*dingaka*) at important times during the process (which can last for up to two years or more) and the giving of formal advice to the couple by older male and female kin (the so-called giving of 'words' on how the couple should treat each other when married).

There is a long history of the anthropological study of marriage in this part of Africa deriving from the early work of Schapera in the 1930s

---

[2] Prestation is an anthropological term developed by Marcel Mauss to indicate a payment that is due yet usually takes the form of a gift, such as a tribute in tributary systems.

and 1940s, for example Comaroff & Roberts (1977), Kuper (1982), Griffiths (1997), Gulbrandsen (1986) and Townsend (1997). These provide a history of the important elements in wedding procedures as well, discussing the increase of *lobola/bogadi* payments, the role of conjugal relations *vis-à-vis* concubinage (the so-called *inyatsi*) and the significance of kinship patterns.

These scholars have emphasized what can be called an ethnicization of marital practices and requirements. Over time and as a result of the increasing involvement of local populations in long-distance labour migration, migrants and their home communities have become interested in emphasizing ethnic differences by making clearer distinctions in the bride-wealth and gift-giving practices between different ethnic groups. In other words, growing contact between members of different ethnic communities (through increasing mobility to South African industrial complexes) seems to have had an impact on the way people began to define the significance of their cultural requirements for marriage.

The Bakwena are one of the eight 'principal tribes of Botswana' and in Molepolole, the older generation consider it important to stress that the standard *lobola* in Bakwena culture is set at eight head of cattle. A further head of cattle is required if a 'jumping of the fence' and a 'spoiling of the yard' takes place, which means a woman becomes pregnant before official recognition of a relationship has taken place (Molokomme 1991). While fulfilment of these requirements always remains subject to agency and negotiation, it is unclear when exactly this idea of a fixed number of cattle and similar requirements became so dominant. Exploring this question through a multi-generational approach in which I not only interviewed the members of the parental generation of the young couples but also their grandparents indicated that the number became fixed in the later years of labour migration to the mines, i.e. in and around the 1950s. Before then, there was, according to my interlocutors, either more space to marry without fulfilling the obligation of a *lobola/bogadi* payment or the chance to exchange a much smaller number of cattle.

In other places and among other ethnicities in Botswana, either the number of head of cattle was and still is different, or the bride wealth can be composed of a broader mix of livestock, such as goats and sheep. Authors such as Townsend (1997) argue that in all cases of *lobola*, long-distance marital connections have remained intact (irrespective of the relations men and women had while being separated) and the residential location of offspring was secured in kinship patterns and *vis-à-vis* inheritance and authority.

The idea of *lobola/bogadi* as a crucial element in wedding arrangements is still highly dominant in Molepolole, yet its adaptation to a context of long-distance migration has almost completely vanished. While in the pre-independence and colonial years it was customary that men and women married within the locality of Molepolole (from one of its many wards into another), in the post-independence years marriages became increasingly neo-local. This was due to the fact that new labour markets were opening up within Botswana resulting from a rapidly expanding educational system that allowed for greater opportunities to meet partners from different parts of the country and from various ethnic backgrounds. While 'ethnic' in terms of requirements, *lobola* payments became increasingly negotiable across ethnic divides. Yet this negotiability does not mean that a marriage without *lobola* has become a recognized option for family elders.

One of the fundamental requirements that the younger generation has to meet when getting married is to ensure that the *lobola* payment fulfils anything that the elders have stipulated in *patlo* negotiations, even if they are marrying across an ethnic divide. There is no greater embarrassment than to be disrespectful in this sense but one of the new constraints is that, unlike in the past, the couples *themselves* have become largely responsible for collecting or buying the required number of cattle. The fundamental shift here is that it is no longer automatic or self-evident that the patriclan of the man is responsible for delivering the required cattle. By and large, this important element of 'tradition' has now been devolved to the couple, such that, couples often claimed that they were very much left to themselves to see where and how they were going to deliver the eight head of cattle. Even if some of these cattle are given to them by relatives, the bride frequently contributes to the payment of her own bride wealth, either directly or indirectly.

While discussing a social cocktail of requirements and their development over time in the context of weddings, the second most important requirement is the tent (van Dijk 2010, 2012). In terms of costs, the tent even outstrips the costs for the payment of *lobola*. Contrary to the *lobola*, there is little sense of a deep-rooted cultural history here but, interestingly, being able to erect a tent at the parental homes of the bride and the groom attributes much public respect to couples and their families (as the *lobola* payment also does) but it also espouses an identity to the couples and their families as being 'modern', if not 'cosmopolitan'. The expensively decorated tents can be seen as a sign of success in the consumerist society in which middle-class styles and appetites have become mainstream. The tents contribute to the colourful aesthetics of the wedding as they make for a splendid sight with decorated tables, flowers, ribbons and other decorations that signal prestige and success. Families pressurize couples into having these tents as they attract attention in the local neighbourhood, make a representative place to welcome important guests and visitors, and decorate the parental homestead in ways that would otherwise never take place. They signal the economic success of the family and these couples, their prosperity and the way in which they have been part of the upward social mobility in Botswana (Mannathoko 1999; Durham 2002; Werbner 2004; Good 2008; Livingston 2009; Dobson 2011).

While there is a history to the relationship between a changing economy and marital practices, these decorations, tents and other cosmopolitan artefacts have become true signs of the new middle class. In collecting marital histories in the 1970s among the Tswana of Mafikeng, Comaroff & Comaroff (1997: 386) also noted, with an element of surprise, how gradually 'weddings became major sites of conspicuous consumption' for the local elites, involving elaborate gowns, invitations, gifts, feasts and photographs to mark of their elevated status. Yet they do not note that this is a form of distinction for which the (economic) responsibility nowadays rests entirely with the couple. They are responsible for saving or borrowing enough money to rent these tents, to throw a big party once the tent is erected and to make sure all their guests have enough to eat and drink. Weddings can be preceded by 'showers' and bachelor parties where the couple will receive gifts in cash and kind from friends, colleagues, fellow church members and the like, and at the so-called after-party they can expect to receive similar gifts. Yet the money received is usually only a minor contribution to the total expenses that the couple will face. Many of the 25 couples that I followed in Molepolole had no other option than to take out a bank loan, in some cases the equivalent of a whole year's salary.

While the tent is one of the most visible elements in the 'middle-classization' of weddings, other elements of the globalization of weddings are just as much a part of the modern outlook of these events. As a growing literature shows, modern mediatized images of marriage and relationships on TV, in glossy magazines and on the Internet are part of the local fascination with expensive styles (Hirsch & Wardlow 2006; Cole & Thomas 2009; Masquelier 2004, 2005). These include objects such as rings, flowers, wedding cards, wedding dresses and outfits for the bride, groom, best man and the maids of honour, cars, decorations, videos, presents, photo shoots, after-parties and loud music. In addition to pressure from parents and other senior kinsmen, peer pressure is important too. Competition is part of the deal as families and couples strive to outdo others, to have the pictures of these events in the glossies or display them in the local community, for example in churches. While weddings have become firmly located in a neo-liberal market of middle-class consumptive styles, of firms that have discovered weddings to be a profitable avenue and moneylenders who have found a niche for profit-making through the provisioning of personal loans, the wedding market is one of fierce competition of all sorts that extends way beyond the domains of social control and the interaction of individual families. In other words, new competences of budgeting, fulfilling the requirements of taking out bank loans, knowing where to look on the Internet for the latest wedding fashions and how to dress and style the event are required to meet the demands of this prestational capitalism in which the emerging middle classes have situated themselves.

A third domain of requirements that this social cocktail entails relates to the formation of Botswana's civil society, the place of marital law, changing patterns of ownership of property, inheritance and family law. Officially and under Roman Dutch law, Botswana recognizes three different systems of matrimony: customary, religious and civil marriage. What usually happens is that couples register their marriage under Roman Dutch law at the office of the District Commissioner (DC). Of the, on average, 300 marriages that are registered in Molepolole every year (van Dijk 2010), fewer than ten are registered under customary law at the Paramount Chief's palace (the *kgotla*). The majority are registered at the DC's office, fulfilling the requirements of Botswana civil law that states that the register of marriages is signed by the couple and their witnesses. There are complexities here as it is never absolutely clear if one of the couple is already married and neither the recording of marriages under customary law at the Paramount Chief's office nor the officiating at weddings by religious officers in Christian or Islamic communities is a water-tight system.

Polygamy is not allowed under Roman Dutch law and Christianity's efforts to stamp it out have been formidable, particularly in Molepolole where one of the first Christian mission stations was set up. Comaroff & Comaroff (1997: 375) noted that one major impact of Christianity in the region has been its emphasis on making 'custom' appear sinful and legality virtuous. The Tswana were expected to develop a wish to marry according to the law, which would thereby narrow the range of 'acceptable' relationships down to legal wedlock exclusively. Over time, this pursuit of legalism has become the sign of respectability, not only defining the material implications in terms of marital rights to property, inheritance and so forth but also creating an 'external memory' of the nature of marital relations. This means that respectability has become highly dependent on the quality of record keeping by the DC, chiefly or religious offices. Such records should demonstrate under which of these matrimonial forms, which in themselves hold statutory power, a certain individual has entered a particular marital relationship so as not to jeopardize the pursuit of the nation-state regarding enforcing monogamy.

In Molepolole, the pattern has been that the less or non-educated have been opting to 'register' their marriage at the Paramount Court, whereas the educated classes have generally registered with the DC. The situation for religious marriages is less clear as Islamic religious law allows for polygamous relations. In Christian circles there has been a tendency to establish so-called marriage officers (usually prominent members of their religious community) who are given the authority to issue a marriage certificate if the marital vows are taken in church.

While Mamdani (1996) assessed political positions of individuals in African states as being bi-focal in terms of being a subject under traditional authority and a citizen under state authority, the complexity of the wedding system in Botswana is such that it is necessary to talk about a tri-focal system. However, irrespective of the different forms of authority that pronounce on the legality of any marital relationship, in the minds of family elders, a marital relationship can only be recognized when the couple has obtained their consent and has performed the rites according to *ngwao*. If there is any conflict about their authority or the required payment of *lobola*, the status of a relationship is referred to the Paramount Court. Yet for the couples in my research sample, many of whom are well educated, have an urban lifestyle, modern job and paid employment, some kind of supremacy by the customary courts over marital relationships that have been signed and registered under Roman Dutch law is difficult.

Since the 1970s, Griffiths (1997) has been exploring complex conflicts and contestations in this domain from a legal-anthropological perspective because of the cocktail of matrimonial forms, jurisprudence and customary and other forms of authority. For many of today's couples, it is important to have their relationship recognized not only in customary terms but also under Roman Dutch law since their property and inheritance rights are best protected under state law in situations of misfortune, sickness, death and divorce. What is termed 'property grabbing' under customary law still happens and often leaves the women destitute if any kind of predicament befalls her.[3] The Botswana nation-state has been interested in promoting modern family law and passed the BAMP Act (the Botswana Abolition of Marital Power Act) in 2005, a law that ended the husband's monopolistic right over a couple's property. While this law is meant to cover the whole of Botswana, its reach into customary and religiously recognized marital relations is uncertain, as is the extent to which the elders adhere to this law in the case of death and divorce.

The fourth domain in considering the complex requirements that this social cocktail of marital relations and practices entails is morality. While there is a long history to the way many forms of Christianity came to be an influential power in the shaping of monogamous marital relations and in the tying of sexuality exclusively to the realm of the matrimonial bond, the complexities are enormous. First of all, Christianity is not one unified whole. When considering Molepolole, it should be noted that some of the so-called mainline Christian traditions have been dominant in the shaping of relations among the Bakwena. Livingstone established one of the first mission posts of the London Missionary Society in the village in an attempt to convert the Bakwena chieftaincy to the faith in the second half of the nineteenth century. The church later developed into the United Congregational Church of South Africa (UCCSA) and has remained one of the largest and most influential churches in Molepolole. Yet the London Missionary Society was followed by other mainline missionary churches, such as the Roman Catholic Church, the Anglican Church, the Lutheran Church, the Methodists, as well as Pentecostal missionary churches, such as the Assemblies of God.

A further step in the development of Christianity was the establishment of so-called African Independent Churches, of which Molepolole today hosts a wide variety. The important congregations here are the various Zion Christian churches, Apostolic churches, spiritual healing churches, even churches such as Eloyi, which were all started under African leadership with a broad repertoire of specific symbolism. In some cases, these churches differed from the mainline churches' emphasis on the fight against polygamy, in others they were becoming even stricter

[3] Property grabbing indicates the context in which, as is customary in Molepolole, the woman marries into her husband's family. In a sense, she remains the stranger and, in the case of divorce, is expected to 'return', i.e. to go back to her paternal home. If her husband dies, customary inheritance notions are such that any property as well as the children born in wedlock remain part and parcel of the patriclan of the husband.

in the control of extramarital relations than was, and still is, the case in the mainline traditions. In the 1970s, African Pentecostal churches began to be established in Molepolole and were strict in terms of their control of marital relations as well as sexuality in terms of premarital and extramarital relations.

This development in Christianity tied itself to the functioning of the nation-state in its moralizing of the marital bond. This is clear, for example, from the way in which the Registrar of Marriages addresses couples with prayers and Bible readings, stressing the Christian principles of monogamy. In a sense, Christianity is part of what can be called a 'civilization offensive' that is intent on making a marital morality part of the well-functioning of civil society. In this process, gender roles are usually also specified, irrespective of reality and the fact that women have had more chances of creating independent positions in society due to education and the growing job market. The Christian marriage has become closely connected to ideas of the 'Christian home', 'Christian courtship' and 'Christian childrearing' (Comaroff & Comaroff 1997).

Over the past two decades, this moralizing agenda regarding marital relations has gained new impetus as a result of the AIDS pandemic. With the rare combination of relative wealth and high rates of AIDS infection, Botswana began to include Christianity in its fight against AIDS by embracing the moralizing capacity of the faith *vis-à-vis* relationships, sexuality and reproduction. Increasingly, the formation of marital relations became subject to a process of marital counselling. Couples, especially those that belong to any of the country's many Christian denominations, are requested – in some cases such as in the Pentecostal churches even pressurized – to have marital counselling before their wedding. Depending on the nature of their Christian denomination, which can vary from relatively 'easy-going' mission churches to African independent churches and highly demanding and rigid Pentecostal churches, this counselling will venture to explore the couple's private lives. In addition to elements of faith and religious conviction, 'behavioural change' is also on the agenda, i.e. promoting 'self-responsibility' by the couple *vis-à-vis* their relationship, sexuality and reproduction (Klaits 2005). In Pentecostal circles, forbidding sex before marriage is taken very seriously and churches stress 'A & B' messages: Abstinence and Being faithful (Nkomazana 2007). While international AIDS campaigns usually stress A, B & C messages (the C being for condomizing, i.e. safe-sex practices), the Pentecostal churches usually omit the 'C' message as condoms are thought to promote playfulness, infidelity and irresponsibility.

In addition to their sexual content, counselling sessions also explore issues such as health and hygiene, past relationships, children from earlier relationships, inheritance rights, relations with the extended family and money matters. While Botswana is generally marked by a culture of utter discretion when it comes to discussing matters between the generations relating to sexuality, reproduction and financial affairs, these counselling sessions may be shockingly confrontational for couples. Although conversations take place in an intimate context in a direct exchange between pastors/counsellors and the couple themselves, the couple is being put in a position where they are expected to discuss intimate matters with a person much older than them, which is a very unnatural situation for them. These sessions deviate from more traditional settings of giving and receiving advice as they present a modernity of relations and a new openness concerning intimate matters.

This is another element in the social cocktail of practices that have, over time, been surfacing in marital practices in Botswana and that are responding to a dramatically changed reality, namely the arrival of AIDS.

The social cocktail of competences: Placing the couples

Couples are being faced with an increasing level and variety of competences that they need to master due to the rising complexity of the marriage process. Perhaps even stronger than the innovations in the marital process as such since Independence are the innovations in the competences themselves, which are now more salient and telling. While in the years of long-distance migration, marital practices in a sense accommodated the reality of the extended absence of spouses, the reality did not require any profound innovation on the part of couples' competences to deal with the situation. Commonly, marriages were arranged by families, the *lobola/bogadi* was part of intricate family reciprocities and the financial burden that rested on the shoulders of the couple themselves would usually be limited to buying customary blankets (worn by older women as a sign of their status) and some nice clothes. The celebration itself would have been modest.

Today, however, there is no simple adaptation of couples' competences for dealing with the rise in wedding costs and all the other legal, moral and social aspects surrounding marriage. These competences came to be part of a profound innovation of personal skills, social brokering, identity management and the processing of information. While all of these innovations serve many other purposes too, such as preparing people to be global citizens, producers, consumers, recipients of public services and custodians of natural resources, marriage is but one of the many challenges facing today's youngsters. As Ferguson (1999) aptly put it, it has become another 'expectation of modernity' that demands to be fulfilled. Couples embarking on the wedding process nowadays have become veritable entrepreneurs and marketing experts of their own personal lives. They need to have the capacity to understand banking systems, medical systems for testing, CD-4 counts, ARV treatment and genetic profiling and should be competent in understanding customary rulings on marital practices and chiefly and elderly authority. Understanding global systems of middle-class lifestyles and having the social skills to negotiate peer pressure, personal pleasure and parental desires while controlling the demands being placed on their personal resources are also vital.

Talking with couples about how this new cocktail of competences, skills, capabilities and resources is being brought together is time-consuming. It is they themselves who usually begin to emphasize how this 'management' has been revolutionizing their personal lives, often to their own astonishment. It has become the single most complicated endeavour of their lives, stretching from before the wedding to dealing with the long-lasting effects of servicing debts and bank loans years after the wedding, building a neo-local home, checking their AIDS status and securing their property. Interestingly, a clear-cut distinction of the different elements in the process that makes for the social cocktail that the wedding is and has become, i.e. a fictive separation of the elements that should belong to 'tradition', civil responsibility, Christianity and middle-class global styling – is completely irrelevant. The competences and capabilities that they possess or should master are to be enacted in any of the contexts that this cocktail entails.

Whether they are talking to their parents about *lobola/bogadi*, to the bank manager about a bank loan, to the pastor regarding the intimate aspects of life or to the District Commissioner to gain information about the legal ins-and-outs of the matrimonial state, they always need to consider their individualized position. A nostalgic view of 'families that marry', i.e. a perspective that emphasizes the relational nature of marriage and wedding arrangements, would deny couples this fundamental experience, namely that they as men and women are required to position themselves individually in the process. 'We are on our own' couples explained to me when reviewing their situation. The cocktail of marriage patterns, requirements and constraints has become such that a nuptial ideology of families that marry their children off to other families and that marriage should be seen as a collective responsibility is a far cry from the reality of highly personalized competences and capabilities that couples must master. Couples are responsible for servicing any bank loan they may take out to finance their wedding and cannot depend on support from their respective families in this matter.

The phrase 'we are on our own' (i.e. that they can expect very little support from their relatives and the realization that any form of romantic ideal of family support is misgiven) is true for both the partners. Interestingly, as the financing of marriages is such that each of the spouses will bring his/her own financial contribution from savings, personal income and personal loans, even the buying of the head of cattle that are required for the payment of the *lobola* has become part of their respective responsibility. Collecting cattle from family members on the men's side (that was a form of reciprocity with regard to the patriclan) is disappearing rapidly as negotiations over cattle turn into discussions about the cash equivalent of eight head of cattle at the cattle market.[4]

While the nostalgic notion of a communal, collective engagement and responsibility for marriage has been replaced by a comprehensive takeover of a market-oriented individualized competence, a romanticized idea of 'free choice' in a passionate relationship is also not present. I am not saying that there cannot be passion or love in relationships that are established through this cocktail of practices, but the notion is much more, i.e. that this is superseded by the pursuit of organizational skills, fulfilling the appropriate roles and obligations at the right time, making no mistakes during the complicated proceedings and ensuring that all the relevant authorities and players in the process are happy about what is happening. Couples explained this to me in terms of being able to keep many balls in the air at the same time and of demonstrating to one another that they are dedicated, committed and adamant about seeing the whole process through. Espousing and acquiring the competences to manage the process is often as much a test of the relationship as it is of one's individual qualities.

The fundamental innovation, more than an adaptation to changing circumstances that this social cocktail is creating, is that there is a simultaneous transformation going on. While gender role models are being

[4] In 2012, the price per head of cattle varied between BP 1,200 and BP 2,000 (US$ 120 to US$ 200).

transformed by the ways in which men and women define their masculine and feminine positions *vis-à-vis* their social position (Suggs 2002), the model of the marital relationship is also changing under the weight of this cocktail. Although male and female role models have never been perceived in isolation but as responding and interacting dialogically with one another, the modelling of the relationship that is called marriage is being reshaped separately. Male and female models in Botswana have changed dramatically because of the country's prosperity and the opportunities offered by the arrival of new middle-class lifestyles, access to education, better-paid jobs, changing laws and women's rights (Dobson 2011; Livingston 2009; Good 2008; Werbner 2004).

While we cannot yet celebrate as levels of inequality still exist between men and women in Botswana, male roles have changed (Suggs 2002). With increasing numbers of educated women accessing high-ranking positions in various sectors of the Botswana economy, ideas about male responsibilities for relationships, sexuality, the home and the sharing of property rights have influenced a younger generation and men are increasingly accepting that women can also take out bank loans and co-sponsor weddings, the buying of a house and/or a (family) car. Women have acquired a greater level of economic independence from their husbands and their parental families as a result of education and the country's rapid economic development.

At the same time, the biggest innovation is the way these role-model changes have come to influence the modelling of marital relations today. The individual competences that are required to make marriage happen have fundamentally altered the relationship model of the younger generation. The four frameworks of these competences that were identified above (i.e. the traditional ceremonial, the modern consumptive, the law, and the moral behavioural framework), which are partly overlapping but in many ways also compete, make for this new mix that is impacting on the lives of many.

In conclusion, weddings today are demonstrating a qualitative difference between adaptation and innovation that may not only imply the aspect of time and social change but also an element of the extent to which a wider social environment is being affected and changed in the process.

References

Appadurai, A. (1996) *Modernity at Large: Cultural Dimensions of Globalization*, Minneapolis: University of Minnesota Press.

Bastian, M. (2001) 'Acadas and Fertilizer Girls: Young Nigerian Women and the Romance of Middle Class Modernity'. In: D. Hodgson, ed., *Gendered Modernities: Ethnographic Perspectives*, New York: Palgrave, pp. 53-76.

Brown, B. (1983) 'The Impact of Male Labour Migration on Women in Botswana', *African Affairs* 28(328): 367-388.

Bruijn, M. de, R. van Dijk & J.-B. Gewald, eds (2008) *Strength beyond Structure: Social and Historical Trajectories of Agency in Africa*, Leiden: Brill.

Cole, J. & L.M. Thomas, eds (2009) *Love in Africa*, Chicago: University of Chicago Press.

Comaroff, J. & J. Comaroff (1997) *Of Revelation and Revolution. Vol. 2, The Dialectics of Modernity on a South African Frontier*, Chicago: University of Chicago Press.

Comaroff, J.L. & S. Roberts (1977) 'Marriage and Extra-Marital Sexuality: The Dialectics of Legal Change among the Kgatla', *Journal of African Law* 1: 99-123.

Constable, N. (2009) 'The Commodification of Intimacy: Marriage, Sex, and Reproductive Labor', *Annual Review of Anthropology* 38: 49-64.

Cornwall, A. (2002) 'Spending Power. Love, Money, and the Reconfiguration of Gender Relations in Ado-Odo, Southwestern Nigeria', *American Ethnologist* 29(4): 963-980.

Dobson, J. (2011) 'New Silhouettes of African Women: How Women in Botswana Juggle Work and Home-Based Roles', *Honors Projects*, Paper 11. http://digitalcommons.macalester.edu/anth_honors/11

Durham, D. (2002) 'Uncertain Citizens: The New Intercalary Subject in Postcolonial Botswana'. In: R. Werbner, ed., *Postcolonial Subjectivities in Africa*, London: Zed Books.

Ferguson, J. (1999) *Expectations of Modernity: Myths and Meanings of Urban Life on the Zambian Copperbelt*, Berkeley/Los Angeles: University of California Press.

Good, K. (2008) *Diamonds, Dispossession and Democracy in Botswana*, Oxford: James Currey.

Griffiths, A.M.O. (1997) *In the Shadow of Marriage: Gender and Justice in an African Community*, Chicago/London: University of Chicago Press.

Gulbrandsen, O. (1986) 'To Marry – or Not to Marry. Marital Strategies and Sexual Relations in Tswana Society', *Ethnos* 15(1/2): 7-28.

Hirsch, J.S. & H. Wardlow (2006) *Modern Loves: The Anthropology of Romantic Courtship and Companionate Marriage*, Ann Arbor: University of Michigan Press.

Johnson-Hanks, J. (2007) 'Women on the Market: Marriage, Consumption and the Internet in Urban Cameroon', *American Ethnologist* 34(4): 642-658.

Klaits, F. (2005) 'The Widow in Blue: Blood and the Morality of Remembering during Botswana's Time of AIDS', *Africa* 75(1): 46-62.

Kuper, A. (1982) *Wives for Cattle: Bridewealth and Marriage in Southern Africa*, London: Routledge & Kegan Paul.

Livingston, J. (2009) 'Suicide, Risk, and Investment in the Heart of the African Miracle', *Cultural Anthropology* 24(4): 652-680.

Mamdani, M. (1996) *Citizen and Subject: Contemporary Africa and the Legacy of Late Colonialism*, Princeton: Princeton University Press.

Mann, K. (1985) *Marrying Well. Marriage, Status and Social Change among the Educated Elite in Colonial Lagos*, Cambridge: Cambridge University Press.

Mannathoko, C. (1999) 'What Does It Mean To Be a Middle-Class Woman in Botswana?' In: C. Zmroczek & P. Mahony, eds, *Women and Social Class: International Feminist Perspectives*, London: University College London, pp. 9-23.

Masquelier, A. (2004) 'How is a Girl to Marry Without a Bed? Weddings, Wealth and Women's Value in an Islamic Town of Niger'. In: W. van Binsbergen & R. van Dijk, eds, *Situating Globality: African Agency in the Appropriation of Global Culture*, Leiden: Brill, pp. 220-256.

—— (2005) 'The Scorpion's Sting. Youth, Marriage and the Struggle for Social Maturity in Niger', *Africa* (11): 59-83.

Molokomme, A. (1991) *Children of the Fence: The Maintenance of Extra-marital Children under Law and Practice in Botswana*, Leiden: African Studies Centre.

Nkomazana, F. (2007) 'Christian Ethics and HIV/AIDS in Botswana'. In: J. Amanze, O.N. Kealotswe & F. Nkomazana, eds, *Christian Ethics and HIV/AIDS in Africa*, Gaborone: Bay Publishing, pp. 28-47.

Pauli, J. (2009) '(Re-)Producing an Elite. Fertility, Marriage and Economic Change in Northwest Namibia'. In: C. Greiner & W. Kokot, eds, *Networks, Resources and Economic Action*, Berlin: Reimer, pp. 303-325.

Schapera, I. (1940) *Married Life in an African Tribe*, London: Faber & Faber.

—— (1950) 'Kinship and Marriage among the Tswana'. In: A.R. Radcliff-Brown & D. Forde, eds, *African Systems of Kinship and Marriage*, London: Oxford University Press, pp. 140-165.

Solway, J.S. (1990) 'Affines and Spouses, Friends and Lovers: The Passing of Polygyny in Botswana', *Journal of Anthropological Research* 46(1): 41-66.

Suggs, D.N. (2002) *A Bagful of Locusts and the Baboon Woman: Constructions of Gender, Change, and Continuity in Botswana*, Belmont, CA: Wadsworth.

Townsend, N.W. (1997) 'Men, Migration, and Households in Botswana: An Exploration of Connections over Time and Space', *Journal of Southern African Studies* 23(3): 405-420.

van Dijk, R. (2004) 'Negotiating Marriage: Questions of Morality and Legitimacy in Ghanaian Pentecostal Diaspora', *Journal of Religion in Africa* 34(4): 438-468.

—— (2010) 'Marriage, Commodification and the Romantic Ethic in Botswana'. In: M. Dekker & R. van Dijk, eds, *Markets of Well-being. Navigating Health and Healing in Africa*, Leiden: Brill, pp. 282-305.

—— (2012) 'A Ritual Connection: Urban Youth Marrying in the Village in Botswana'. In: M. de Bruijn & R. van Dijk, eds, *The Social Life of Connectivity in Africa*, New York: Palgrave/Macmillan.

Werbner, R. (2004) *Reasonable Radicals and Citizenship in Botswana. The Public Anthropology of Kalanga Elites*, Bloomington: Indiana University Press.

11

Of labradors and libraries: The transformation of innovation on a farm in Kibale, western Uganda

Jan-Bart Gewald

Focusing on a former colonial residence that was transformed into a tourist guest house on a tea estate and farm, this chapter describes how innovation, developed and acquired from around the globe, came to be appropriated, transformed and applied in western Uganda. Established by the colonial administrator, C.W. Switzer, Isunga Tea Estate drew upon a wide variety of sources ranging from African labour to South Asian agricultural policies and Oxbridge norms and values to ensure their success. It is argued that any attempt to understand Isunga Tea Estate will need to take into account the innovations, developed elsewhere, that were appropriated, transformed and applied in Uganda and, in the process, have become uniquely Ugandan.

Introduction

In a lecture given in London in the early 1960s entitled 'In Search of Cultural History', the Austrian art historian, E.H. Gombrich, speaking in part as a *Bildungsburger* of the former Austro-Hungarian Empire, discussed Beethoven's Pastoral Symphony, noting that, although he himself had never personally 'seen merry-making peasants quite in the style Beethoven saw (them)', the pastoral did evoke in him 'clusters of cultural resonance' that did not require any explanation. However, time had relentlessly marched on to 'ribbon developments (and) . . . petrol stations' to a period when 'there are no merry-making peasants either in life or even in fiction'. Reflecting on this, Gombrich (1969: 45) warned that 'our own past is moving away from us at frightening speed, and if we want to keep open the lines of communication which permit us to understand the greatest creations of mankind we must study and teach the history of culture more deeply and more intensely than was necessary a generation ago, when many more of such resonances were still to be expected as a matter of course'. His innate and immediate understanding and appreciation of the 'cultural resonances' in Beethoven's Pastoral Symphony are akin to my experience of the 'cultural resonance and residue' that the British Empire evoked in me during my visit to the Chimpanzee Forest Guest House in Kibale in western Uganda. Much as Gombrich's merry-making peasants had ceased to exist, the colonial officials of the Empire have largely passed away, yet their legacy continues to resonate for those schooled and attuned enough to see and feel it. Although nowhere near as cultured or inspiring as Beethoven and Gombrich, it is in an attempt to capture the 'cultural resonances' of the British Empire, which the Chimpanzee Forest Guest House evoked in me, that this chapter has been written.

The chapter highlights a few phenomena observed at the Chimpanzee Forest Guest House that evoked the British Empire in my mind and seeks to show how these have come to be transformed in their consumption and appropriation within western Uganda. As such, the chapter is shamelessly an exploratory piece that seeks to put into words the cultural resonances of the British Empire that I so clearly felt when I stayed at the guest house in Kibale. The chapter has partly been inspired by the work of Neil MacGregor (2010) who, as curator of the British Museum, selected 100 objects from the museum's collection to frame and write *A History of the World in 100 Objects*. However, it does not deal solely with the material detritus of the British Empire but is far more interested in the ideas and practises associated with the British Empire that have remained and been transformed. It follows the work of John M. MacKenzie who, as series editor for Manchester University Press's 'Studies in Imperialism', pioneered the study of cultural and popular imperialism, with 'the conviction that imperialism as a cultural phenomenon had as significant an effect on the dominant as on the subordinate societies'.[1] It is MacKenzie's eye for culture that is particularly relevant here and his willingness to illustrate, explore and explain the British Empire through more than merely politics, military might or social and economic forces (see MacKenzie 1984, 1986,

[1] http://www.dalmackie.com/main_v2010_Sep.swf. Publicity website of the 'Studies in Imperialism' series, edited by John MacKenzie. Accessed 14 September 2012.

1992, 2001a, 2011b; Richards & Mackenzie 1986). In dealing with the cultural encounter and translation, that the British Empire was, MacKenzie's work echoes, illustrates and does justice to the earlier work by Gombrich (1969: 41) who, in a wholly different context noted:

> ... the difference between the social and the cultural historian. The first is interested in social change as such. He will use the tools of demography and statistics to map out the transformations in the organization of society. The latter will be grateful for all the information he can glean from such research, but the direction of his interest will still be in the way these changes interacted with other aspects of culture. He will be less interested, for example, in the economic and social causes of urban development than in the changing connotations of words such as 'urbane' or 'suburbia' or, conversely, in the significance of the 'rustic' order in architecture.

In keeping with Gombrich and MacKenzie, this chapter is interested in the cultural encounter and cultural translation, both in the past and the present, of the British Empire and what it entailed for western Uganda.

In Germany, following the efforts primarily of Alf Lüdtke's *Alltagsgeschichte* (everyday history), there has been a plethora of work bringing to the fore hitherto hidden histories of the German working classes. *Alltagsgeschichte* was characterized by its innovative use of sources that would otherwise not have been used for historical research (Crew 1989). In particular, Lüdtke and his students used sources and methods akin to those used as a matter of course within African history. Historians of Africa, in contrast to students of twentieth-century German history, have a more limited palette of written (published and archival) sources to draw from. Consequently, those choosing to study African history have developed an eclectic methodology and epistemology that sees history in a wide and varied array of sources. Drawing on anthropological and sociological research and methodologies, they find history in a broad and never-ending range of places. Sigfried Giedion's position (cited in Burke 2010: 75) that 'for the historian there are no banal things' since 'tools and objects are outgrowths of fundamental attitudes to the world' certainly holds true for Africa. Much as Norbert Elias was able to fruitfully explain social structure in the seemingly eccentric topic of table manners, so too historians of Africa have found fertile research in topics as varied as 'the handshake' or 'the top hat' (Ross 1990, 1999: xii, 203).

Although the work of the French sociologist Michel de Certeau developed in a wholly different context, his work parallels that of Robert Ross and is useful when thinking about African History. De Certeau emphasized the creativity and inventiveness of people in their consumption of mass-produced items, be they television programmes or televisions, and analysed the practices of ordinary people, 'everyday practises such as shopping, walking a neighbourhood, arranging the furniture or watching television' (Burke 2010: 80). De Certeau showed that people as consumers were creative in their consumption, that there was no single predetermined form of consumption and that consumption should be seen as a form of production. In this, de Certeau parallels the work of Social Constructivism (SCOT) theorists, although unlike them, he had an eye for the immaterial.

Science and Technology Studies (STS) have taken off particularly in the light of the Social Constructivism (SCOT) approach, which stands in stark contrast to technological determinism and emphasizes the fact that technology development is primarily a social process. Bijker (2001) has noted that within STS there has been particular interest in the social aspects of design and development processes, and far less in the role of users. Mirroring de Certeau's work, Kline, Pinch and Oudshoorn have indicated that the role of users goes further than merely the adoption and use of technology (see Kline & Pinch 1993, 1996; Oudshoorn & Pinch 2003). SCOT emphasizes that technology has a certain amount of 'interpretative flexibility', which varies within and between social groups: 'The same artefact can mean different things to different social groups of users' (Kline & Pinch 1996: 767). Indeed technology only gains meaning when users make use of the possibilities that are contained within its particular forms. In keeping with Kline & Pinch, this chapter is interested in the manner in which the 'interpretative flexibility of an artefact can reappear at the use stage of technology' and pays particular consideration to 'the relationships between social groups and their ability to shape the development of an artefact and how they in turn get shaped in using it' (*Ibid.*).

In contrast though to Kline & Pinch, and in keeping with de Certeau who did not limit his work to the material, this chapter is not only interested in the material artefacts of the British Empire. Instead, de Certeau's discussion of the 'uses', 'appropriation' and especially of 're-employment' that people made of the goods and practices acquired in their encounters with the British Empire is particularly relevant. As Burke (2010: 80) put it, '(de Certeau) was thinking in terms of ordinary people making selections from a repertory, making new combinations between what they select and, not least, placing what they had appropriated in new contexts'. In other words, people in western Uganda were interested in and selected discrete items from the British Empire, both physical and metaphysical, rather than the structures in which these items were originally embedded as a whole. In appropriating and re-employing the items, people transformed their meaning and use, and it is this transformation that this chapter attempts to illustrate.

On the grounds of a tea estate and small farm in western Uganda near Fort Portal in the foothills of the Ruwenzori Mountains, a guest house consisting of small, spotlessly clean chalets and a campsite caters to the needs of well-heeled tourists, Western backpackers and scruffy overlanders en route to the rainforests of Kibale National Park where they hope to see wild chimpanzees. Guests visiting the estate drive down an avenue of jacaranda trees and are presented with a breathtaking view and frisky dogs that clearly have Labrador Retriever in their ancestry. Situated on a hill, the bougainvillea-bedecked guest house faces north, looking across rolling lawns and tea estates to a lake in a crater in the valley below. To the east, it overlooks tea estates that extend as far as the Kibale National Park, to the south is the summit of the hill and the highest point of the estate, whilst to the west the guest house looks out over valleys and hills covered with small holdings as far as the eye can behold and on to the

Photo 11.1 Chimpanzee Forest Guest House chalet, Kibale [Photo: Jan-Bart Gewald]

snow-capped mountains of the Ruwenzori, Ptolemy's 'Mountains of the Moon'.

Staying at the Chimpanzee Forest Guest House is a liminal experience for anybody who has grown up with the remnants of the British Empire. One is constantly assailed by familiar smells, architectural foibles, table settings, the aroma of certain foods, boisterous dogs, signs painted on doors or tacked to trees, and much more that recalls images from a distant and suppressed past with hedgerows that would not look out of place in the English counties and persistent efforts to recreate England's 'green and pleasant land'.[2]

In itself there is nothing unique in a guest house catering to the needs of holidaymakers desperately seeking to visit and view the last remaining remnants of an imagined ideal and untouched Africa. What is unique though are the historical roots that have led to the development of this successful and well-patronized establishment.

To fully understand the Isunga Tea Estate and the Chimpanzee Forest Guest House, one needs to turn to history, global history in particular. It is only by turning to the global origins of much of what is to be found on the estate and at the guest house that one can truly gain insight into the transformation of innovation that has taken place here. The estate and its guest house are not institutions existing in a timeless void, they exist within a historical flow where ideas and innovations are consistently moulded and changed by society as time passes. Ideas and innovations developed elsewhere evolve and develop in different ways in different settings. Much as the English language has developed variants around the globe that are at times no longer mutually intelligible, so too eating, clothing, building or working styles are continuously evolving in ways that can be independent of one and other. One man's delicious gravy may become another man's inedible brown sludge, one man's highly prized library is an assorted jumble of illegible papers to another, and a model farm is someone else's assault on nature.

This chapter explores the transformation of innovation in Africa through a short analysis of artefacts, behaviour, animals and plants that are to be found on a farm in western Uganda. The first two sections provide the initial geographical and historical setting of the Isunga Tea Estate and Chimpanzee Forest Guest House with a short overview of the development of the Kingdom of Toro and a biographical sketch of C.W. Switzer,

---

[2] Taken from the poem by William Blake, *And did those feet in ancient time*. See also Barfoot (2004: 57).

the founder of the estate. The chapter then moves on to consider food and drink, flora and fauna, and technology, and indicates how these topics contain issues that indicate the transformation of innovation, whereby innovations (both material and cultural) initiated outside of Africa have been transformed within the context of this Ugandan estate.

The Kingdom of Toro

Situated in southwestern Uganda, the Kingdom of Toro is a classic example of transformed innovation as described by Hobsbawm & Ranger (1983) in their seminal work on invented tradition. The Kingdom of Toro developed 'embryonically and precariously, for some fifty years in the middle of the nineteenth century and was extinguished in the late 1870s ... (only to be revived) some ten or a dozen years later due to the initiative of an adventurous Englishman and the opportunism of the son of a former ruler (Kasagama)' (Ingham 1975: 3). With the arrival of British colonial rule in the late 1800s in the form of the Imperial British East Africa Company, Chief Kasagama (Rukirabasaija Daudi Kasagama Kyebambe III) was able to convince Lord Lugard and others that the origins of his authority and his kingdom were traditional, thereby 'satisfying the requirements of "indirect rule" at a time when this ill-defined concept served both as the shibboleth and the escape clause for an overstretched British colonial administration' (*Ibid.*: dust cover). With Kabarole as his capital, Kasagama was fortunate in having the British establish a fort, Fort Portal, in the immediate vicinity of his seat of power (*Ibid.*: 76). Taking advantage of the protection offered by the British, Kasagama and his successor 'enveloped themselves in an aura of neo-traditional authority and surrounded their kingdom with the sanctity of a pseudo antiquity' (*Ibid.*: 3-4). In 1896, following Kasagama's conversion to Christianity, Bishop Tucker and Rev. A.B. Fisher established themselves in Kabarole. In the years that followed, Kasagama's immediate family and followers converted, and his descendents were given the names of British royals, namely, Victoria, Albert, George and Edward. Missionary Broadwood Johnson (1912: 55), who travelled to Kabarole in 1902, described King Daudi Kasagama as being like the British schoolboy comic character Billy Bunter:

> Though about thirty years of age, he is just like a great, burly, rollicking, overgrown schoolboy, taking possession of your heart by the affable geniality and frankness of his manner, and yet making you feel that there is at the same time a dignity of bearing worthy of the first gentleman in the kingdom.

King Kasagama died under suspicious circumstances that were suggestive of poisoning whilst travelling through his realm and was succeeded by his son George Rukidi in 1928 (Ingham 1971: 178).[3] King George Rukidi's brother was named Switzer and it is more than likely that he was named after the colonial official C.W. Switzer.

C.W. Switzer

Kibale Chimpanzee Forest Guest House was founded on the spot where a house and farm stood that had been established by a British colonial officer, C.W. Switzer, the one-time District Commissioner of Kibale. Evidence available at Kibale Guest House indicates that Switzer studied at Oxford as a slightly water-damaged lithograph of Tom Tower in Oxford by J.W. King, dating from around 1920, adorns the sitting room.[4] An entry in the index cards of all those who graduated from Oxford University before 1932 mentions a Christopher Wilfrid Switzer who was born in Hitchin, Hertfordshire on 9 January 1898[5] and was the only son of Bamlet Neptune Switzer, a Clerk in Holy Orders. Before going up to Oxford, he attended Merchant Taylors' School and his home address is listed as St Luke's Vicarage, Bow, London.[6]

---

[3] Rukidi's daughter, Princess Elizabeth of Toro (1983: 6) describes how 'My father, . . . had five wives, but was nevertheless a Christian. . . . he was a good king to his people, who looked on him with love and reverence, for kingship in Africa is, as it once was in Europe, divine'. During World War I, Rukidi served in the King's African Rifles, where he rose to the rank of lieutenant.

[4] Tom Tower, designed by Christopher Wren and built in 1681/1682, forms the main entrance to Christ Church. The tower houses the Great Tom, an enormous bell that chimed 101 times at 21:05 to announce the start of the curfew for the college's 101 students when the tower was completed. The online catalogue of art dealer Sanders of Oxford indicates that J.W. King made two sketches of the Tom Tower, one from the quad and one from St Aldate's. Both sketches date from 1900. http://www.sandersofoxford.com (accessed 8 September 2011). In contrast, http://travellingartgallery.com (accessed 8 September 2011) lists a sketch of the Tom Tower from St Aldate's that is identical to the one on offer by Sanders and dates from 1920. http://www.edinphoto.org.uk (accessed 8 September 2011) indicates that J.W. King was born in London in 1868. The 1891 census has him as a lithographic artist living with his parents in Islington. In the 1901 census he appears in Abingdon, Berkshire, as an artist and frame maker. He is said to have died in Scarborough in 1935. Information available online indicates that King made a number of sketches in Oxford (including one of Magdalen College), Edinburgh and King's College in Cambridge.

[5] Although shipping records at the British National Archives show that a C.W. Switzer who travelled back and forth between Britain and East Africa was born in 1897.

[6] Email correspondence with Judith Curthoys at the archives of Christ Church College, Oxford, 9 September 2011.

War Office files mention a C.W. Switzer who served between 1915 and 1920.[7] Indeed the Oxford Roll of Service has the following entry for Christopher: Switzer, C. W. (Sept. 25 1915) Lt/ 2nd The Buffs (E. Kent Regt.) France 1916; Salonika, 1917. The *London Gazette* of 24 September 1915 (p. 9445) mentions that Christopher Wilfrid Switzer, an ex-cadet of the Officers' Training Corps, was to be promoted to the rank of second lieutenant (on probation) in the 2nd Battalion the East Kent Regiment (the Buffs) at the age of 17 years. Following the second battle of Ypres in May 1915, Britain was desperately short of officers and hundreds of junior officers were rapidly promoted. War Office files indicate that Switzer was awarded service medals for bravery in France and Salonika, and ended the war with the rank of lieutenant.[8] After the war Switzer went to St John's College, Oxford and matriculated on 26 April 1919. He also took a short course in Theology, completing it in 1920, and was awarded a BA on 16 June 1921.[9]

Photo 11.2 Portrait of C.W. Switzer in Kibale [Photo: Jan-Bart Gewald]

7 British National Archives (BNA), War Office (WO) 339/42321, Switzer, C.W. Lieut.

8 BNA, WO 372/19, War Office: Service Medal and Award Rolls Index, First World War.

9 Email correspondence with Judith Curthoys at the archives of Christ Church College, Oxford, 9 September 2011.

In 1925 C.W. Switzer was appointed as a pilot in No. 9 Squadron, a Royal Air Force bomber squadron based at Manston that flew Vickers Virginia aircraft. War Office files mention C.W. Switzer as a lieutenant in 1929 and in January 1930 C.W. Switzer was transferred from Class A to Class C, as a reserve air force officer.[10] It is probable that this transfer was linked to his employment as a colonial administrator. After what was probably his first four-year tour of duty, shipping records indicate that he returned to England and sailed from London to Mombasa in 1934. These records mention two further trips from London to Mombasa in 1937 and 1948.[11, 12]

Due to the absence of research at the National Archives of Uganda, it is not known exactly when Switzer came to be appointed to a position in the Kingdom of Toro in western Uganda, but given that one of the Toro royals was baptized as Switzer, it must have been in the early 1930s during what would have been his first tour of duty (Princess Elizabeth of Toro 1983: 59, 68).

True to form, as an educated Oxford graduate and a colonial official with scholarly ambitions, Switzer published a short piece, 'The Story of Lwabinumi' in the *Uganda Journal* of 1937 noting that: 'I heard this story sung by a minstrel by the camp fire when I was on *safari* in Ruchiga' (Switzer 1937: 262). This substantiates the claim that Switzer was indeed working in southwestern Uganda in the first half of the 1930s. Unlike his colleagues in settler colonies such as Kenya and Rhodesia, Switzer explicitly thanked specific African individuals by name for having collected and made the initial translation of the history that so fascinated him. Switzer unambiguously expressed his surprise that 'a tribe until recently untouched by outside influences' could have composed a story that bore such a striking resemblance to what was to be found elsewhere in the world. Switzer's classical training must have caused him to look beyond initial impressions of 'the most primitive of the tribes' and to see parallels with his own myths of origin. He noted that 'the hero, Lwabinumi, appears to have been the product of parthenogenesis, to have performed

---

10 *Flight*, 7 February 1930: 199.

11 A search of 'Passenger Lists Leaving UK 1890-1960' on the website http://www.findmypast.co.uk (accessed 21 August 2012) lists three voyages by C.W. Switzer from London to Mombasa, Kenya.

12 It is not likely that C.W. Switzer is the same person as K.W. Switzer who was responsible for the loss of two hut tax receipt books at Zomba in Malawi (9 October 1929 to 28 January 1930): BNA, Colonial Office (CO) 525/132/12, Loss of two hut receipt books at Zomba). The files note that an agreement was reached with this Switzer whereby he would pay for half of the cash lost with the receipt books.

many marvellous deeds, to have undertaken difficult tasks in order to win his bride, and finally to have suffered a violent death, but to have been reincarnated in his son' (*Ibid.*). Little did he realize that in his later years he himself would come to exist in popular memory as a mythical founding father in western Uganda. A rather garbled article[13] in Uganda's *Sunday Vision* claims that Switzer adopted the illegitimate children of the English explorer John Hanning Speke, and served as a district commissioner (DC) in Kenya. It is highly unlikely that Switzer ever served as a DC in Kenya, and even more unlikely that he would have adopted the illegitimate children of Speke nearly 100 years after Speke visited the area. However by invoking Speke, the article does provide a theory of origin for people of mixed descent in southwestern Uganda and ascribes stature and magnanimous behaviour to Switzer.

It appears that Switzer was not always stationed in southwestern Uganda. Forester Fred Hummel (2001: 58) makes mention of C.W. Switzer as one of 'only seven British officials stationed at Gulu' in northern Uganda in 1944 where he then served as an assistant to District Commissioner J.W. Steil. It is clear that by the 1950s, C.W. Switzer had married a Toro woman and established himself as a tea planter on the Isunga Tea Estate on the edge of Kibale National Park where his descendants continue to run the estate and the Chimpanzee Forest Guest House.

Food and drink

Although the contributors to the fifth edition of the *Bradt Travel Guide* for Uganda expressed the belief that the culinary expertise of the Chimpanzee Forest Guest House could be better, the food and drink on offer is more than adequate and actually quite good for anyone who has ever experienced boarding-school fare. However, it is the content and presentation of the food and drink served at the guest house that is of interest here. Meals and refreshments are produced and served in a manner deriving directly from English culinary norms and values of the 1950s. Three full meals a day and morning and afternoon tea are served at set times, and meals are served according to a strict table setting with placemats, knives, forks and cutlery for separate courses. In addition to the usual salt and pepper, two additional condiments are always served with every meal that not only speak to the Imperial expansion of England but that, together with

13 'Unearthing John Speke's Lost Seed', *Sunday Vision*, Thursday 8 September 2011. http://www.sundayvision.co.ug. Accessed 8 September 2011.

Coleman's mustard, form the condimental epitome of the British Empire, namely HP sauce and Lea and Perrins Worcestershire sauce.

In the early 1950s, American anthropologist Laura Bohannan conducted anthropological research in Nigeria. In the fictionalized account of her fieldwork, she described the horrors of an English breakfast for an American who would prefer only coffee and fruit, drawing attention to the eating habits and the training of cooks in the British Empire:

> The English like a large breakfast. Indeed, in a country like theirs, where people apply the principles of insulation and central heating to themselves rather than to their houses, a large and greasy breakfast has a certain functional value which disappears in the tropics. The Englishman abroad maintains his traditions tenaciously; he drills them into his servants. I had asked for trained servants; I had, perforce, British-trained servants. (Smith Brown 1964: 13)

Bohannan would go on to experience the delights of boiled tinned sardines for breakfast and other culinary highlights but, as the myriad of entries in the letters and diaries of colonial officials and their dependents demonstrate, the importance of cooks and cooking in the Empire in Africa is not something to be overlooked lightly. Good cooks were literally worth a fortune, ruled over a retinue of assistants and commanded wages far in excess of most other colonial employees. Theodore Williams, writing to his family in 1912, reported how one of his fellow colonial officials had recommended 'a prospective boy – (valet, cook, washer, general slave) who I have not yet booked as he is in spite of the good reports of him very expensive – i.e. 1 lb a month & and 10/- extra a month if he feeds himself'.[14] Twenty-five years later, John Walter, writing to his wife shortly after his arrival in Northern Rhodesia as a colonial officer cadet, illustrated the wage differences between cooks and others when he noted: 'The cook and house boys are cheap (15/- & 16/- per month) & willing, but require looking after, ... Kitchen boy at 5/- ... Garden boy at 7/6.'[15]

Finding a good cook was not an easy matter. Young colonial officer F. Melland, who developed a penchant for thrashing his servants, which certainly did not help his position as an employer, demonstrated his own class background in a letter home when he complained, 'though it's a trivial and common occurrence, to show that one is here tied down almost

---

[14] Rhodes House, Theodore Williams, Mss. Afr. S. 776 – 781. December 5th Friday 1912, Livingstone, My dear mother.

[15] Rhodes House, Walter (John C.W.), Mss. Afr. S. 664. Letters as cadet and District Officer in Northern Rhodesia, written to his wife? 1937 – 39, Isoka, 1 September 1937.

as much as at home by the difficulty of getting good servants especially good cooks'.[16] The absence of good cooks was an issue that spread quickly through the colonial community and had repercussions for the standing of those unfortunate enough not to have one. In the late 1920s, S.R. Denny illustrated this point in a short description of a breakfast with colleagues in Kasempa, Northern Rhodesia: 'It was rather an hysterical meal, because they have just got a new cook and boy, both quite hopeless, and poor Mrs. A. was nearly in tears'.[17] Not only was there the difficulty of finding good cooks, there was also the matter of their trustworthiness. Writing in the Boys' Own style that was all their own, a colonial probationer wrote to his mother, 'We are both pleased with our boys – the cooks are definitely good. Though we shan't be able to tell how much they steal until we get to our stations!'[18] However, good cooks could turn out masterpieces that cheered their employers immensely and caused them to wax lyrical:

> My staff, especially the cook, is improving greatly on acquaintance. The cook is really A.1. – if this is his normal form that he is just finding- he has made me a fine sponge cake today – & turns out custards, fowl & eggs in all their forms, & pie, with huge aplomb. At first he seemed a . . . dolt and did nothing but boil fowls till they were tasteless. I have a suspicion however that he is getting a good few tips from Mbalachi my butler who was butler in a lady's house before and is a smart enough chap to have observed some cooking too.[19]

Losing a good cook was an issue that merited comment: 'I was awfully sorry to lose him, as he was far away the best native cook I had seen in the whole country or in the BCA . . .'.[20] Having no cook whatsoever and being reduced to doing their own cooking was as low as a colonial official could sink, and merited censure.[21]

16 Rhodes House, Mss.Afr.R. 192. Melland (Frank Hulme) Diary of life in northeastern Rhodesia, 1901 – 4 3 vols. August 1901, p. 38.

17 Rhodes House, S.R. Denny, MSS. Afr. S. 791 (2), Subakanya Village, (About 10 m. N.E. of Kasempa), 7th November, 1929. Diary.

18 Rhodes House, MSS. Afr. S. 393. Law (Patrick John) Letters to his parents during colonial service in Northern Rhodesia 1932 – 36. Senanga, Zambesi River, Barotseland, Sept 7th 1932, My dear Father.

19 Rhodes House, Theodore Williams, Mss. Afr. S. 776 – 781. Mporokoso, June 1st 15, My dear father.

20 Rhodes House, Mss.Afr.R. 192. Melland (Frank Hulme) Diary of life in north Eastern Rhodesia, 1901 – 4 3 vols. 11th Jan. 1902.

21 National Archives Zambia, Lusaka, KSE 4/1, The District Note Book Mwinilunga, 'Major Hodson B.N.P, who was sent up in June 1909 to enquire into Mr. MacGregor's administration found him and his assistant, Mr. J.M. Pound, doing their own cooking and

Writing on food culture, Leong-Salobir (2011: 14) has noted that in the period of rapid social transformation in nineteenth-century Britain, new dietary practices evolved in which, 'the choice and preparation of foods, meal times and order of service were socially important and defined class demarcation'. The food habits of the affluent upper classes in Victorian and later Edwardian England became the example that others in the Empire sought to emulate. Canadian author Harold Rhenisch (2006: 9), in seeking to understand his own youth, wrote of school children throughout the British Empire 'from Cape Town to Nairobi to Peshawar, from Singapore to Halifax' who all shared the similar experience of an Imperial education in which 'we ate English breakfasts of eggs, sausage, bacon, and toast' along with Worcestershire sauce, '(which) none of us could pronounce' and HP Sauce.[22]

Across the length and breadth of the British Empire, Worcestershire sauce made the unpalatable palatable.[23] Worcestershire sauce is closely associated with and indeed actually depended on the Empire for its very existence, as the following indicates:

> Despite its name, Worcester sauce was originally an Indian recipe, brought back to Britain by Lord Marcus Sandys, ex-Governor of Bengal. One day in 1835 he appeared in the prospering chemist's emporium of John Lea and William Perrins in Broad Street, Worcester, and asked them to make up a batch of sauce from his recipe. This was done, but the resulting fiery mixture almost blew the head off Messrs Lea and Perrins, and a barrel they had made for themselves was consigned to the cellars. Much later, in the midst of a spring clean, they came across the barrel and decided to taste it again before throwing it out. Wonder of wonders, the mixture had mellowed into a superlative sauce! The recipe was hastily bought from Lord Sandys and in 1838 Britain's most famous commercial sauce was launched. (Burton cited in Porter 2004: 35)

Allowing for undoubted romanticization, it is clear that Worcestershire sauce is a product of the Empire with its many ingredients – malt vinegar from England, tamarind from India, anchovies from the Mediterranean,

washing their own pots and pans – Both officials were recalled, and Mr. MacGregor's resignation asked for.'

22 Following the Unilateral Declaration of Independence in Rhodesia, sanctions were imposed upon the former British colony whereby, amongst other things, imports of HP Sauce came to an end. Instead a locally produced brown sauce in a similarly shaped bottle but with a green label and depicting the Victoria Falls was brought on to the market.

23 Historian Neil Parsons has suggested writing a history of the British Empire through the lens of Lee and Perrins Worcestershire sauce. On its worldwide distribution in archaeological settings, see Hoffmann (2007: 7).

molasses from Caribbean sugar cane – all mixed and matured to cater for the evolving culinary tastes of administrators and settlers around the world. Culinary tastes expanded as connections within the Empire grew and developed and, writing about colonial Asia, Leong-Salobir (2011: 14) noted developments that also applied to East Africa:

> Fruit orchards and vegetable gardens expanded and processed foods became popular. Food manufacturers processed new and exotic foods. Crosse & Blackwell manufactured about 40 different pickles and sauces, Colman's mustard went on the market... The variety of foodstuffs increased even more as more produce from the colonies was brought back to Britain. In addition, food distribution was made easier with the development of the railway system.

Within Britain itself, with the exception of Anglicized sauces that 'disguised... (and) mellowed' dishes, Eastern foods made little impact. Culinary change came about in Britain following the end of Empire and the move of former colonial subjects to the British Isles from India, Kenya and the West Indies. It was the bland fare of boiled potatoes and other vegetables, stewed meat and thick gravy served with Worcestershire and HP sauce (the flash and sparkle of the Empire) that became the standard for British cuisine until the early 1970s. However, even as British cuisine changed, tastes were tempered to suit the still bland British palate. Indeed it has been noted, and as anyone who has eaten Chicken Tikka Masala in Britain will affirm, 'curries in Britain are a post-colonial phenomena (and still usually adapted to British taste)' (Porter 2004: 35).[24]

With their strict table settings, set meal and tea times, gravy boats and Imperial sauces, the food and drink served to the guests at the Chimpanzee Forest Guest House are more than merely an attempt to cater to the tastes of Western tourists. If that were really the case, the kitchen would have been serving hamburgers, French fires and coleslaw, which, thankfully, it does not. Instead the food served by the kitchen, apart from specific local dishes, is food that is considered to be fitting and proper for guests, Ugandan or otherwise. It is food that happens to be based on a culinary tradition that was introduced into western Uganda during the colonial era and is not a failed copy of food served elsewhere. Instead it is a culinary tradition that has taken root and developed over time in rather unexpected ways.

---

[24] Chicken Tikka Masala is a fine example of British cuisine, with the Masala sauce being added to satisfy the British desire to have their meat served in gravy.

Flora and fauna

Isunga Tea Estate and the Chimpanzee Forest Guest House are on the edge of the Kibale National Park, a biodiversity hotspot that, apart from large numbers of mammal species, is home to no fewer than 325 distinct bird species (Awe 2011: 5). The presence of wild chimpanzees and primates is a major tourist attraction in the area and, as the name indicates, the prime attraction of the Chimpanzee Forest Guest House. Kibale Forest has 250 distinct tree species and more than 312 recorded plant species (*Ibid.*: 58). Not surprisingly, the area's excellent climate, with an average annual rainfall of over 1000 mm, and its fertile volcanic soils have led to the intensive settlement of small-scale agriculturalists and large-scale tea estates. A recent impact assessment report put it as follows:

> Salient features of this ecological zone are the luxuriant vegetation, extensive tea estates, and intensive coffee and banana gardens. Other annual crops and small scale livestock rearing takes place in the relatively well to do households. (*Ibid.*: 57)

In short, the Isunga Tea Estate and the grounds of the Chimpanzee Forest Guest House are set in a magnificent natural setting.

Labradors

Dogs as personal pets, as opposed to dogs kept for hunting, racing or guarding purposes, formed an integral part of the social and cultural life of British colonial administrators in the Empire. In Britain itself, dogs proclaimed social standing. In his work on the British middle classes, James (2006: 370) noted that 'the late-Victorian canine hierarchy was one of the many visible reminders that how one spent one's money was, as ever, a determinant of social status'. Ritvo (1986: 227), who has written extensively on domestication and animal breeding in Britain, describes how the demand for purebred dogs as domestic companions snowballed between 1865 and 1887, as 'any other kind of dog might compromise its owner's social status'. In addition, she cites a breeder of pedigree dogs in 1896 who stated that: 'Nobody who is anybody can afford to be followed about by a mongrel dog'.

Access to pedigree dogs in Africa was problematic as more often than not these canines were unable to deal with the rigours of the climate to which they came to be subjected.[25] Native Commissioner Theodore

[25] For a fascinating and entertaining introduction to dogs in African history, see van Sittert & Swart (2008).

Photo 11.3 Chimpanzee Forest Guest House Labrador descendant [Photo: Jan-Bart Gewald]

Williams described employing four carriers in a failed attempt to take two 'Pointer pups' in a Tsetse-fly-proof mesh cage to Avery Jones, District Commissioner of Kawambwa.[26] More often than not, these elaborate and expensive precautions failed and the pedigree dogs died (Swart 2008). Not surprisingly therefore, the successful ownership of a pedigree dog was something that evoked comment from other colonial officials. The majority of colonial officials, however, acquiesced to the situation and acquired locally bred dogs that were generally resistant to the tropical maladies of Africa. As such, the young probationer, Patrick Law, wrote to his father in 1932:

> We have bought a *native* (scratched out) dog since I last wrote. She's a 2 month old puppy of uncertain breed, but we were both so attracted by her that we bought her between us for 2/6, and [when] we part we are going to toss for her. She will I think grow into something like a lurcher. At the moment she's about the size of a small border and a very nice brindled colour. The boys call her Setunda, after the village where we got her.[27]

The dogs at Kibale and the Chimpanzee Forest Guest House are derived from black Labrador Retrievers. Labradors are hunting dogs originally from Canada, and are currently one of the most popular breeds in the world. Hardy, easily trained and with a pleasant disposition, they were much loved by colonial officials. A colonial official who travelled through western Uganda in the early 1960s might well have been describing one of the

---

[26] RH, MSS. Afr. S. 776 – 781.Williams (Theodore R.) Administrative Officer, Northern Rhodesia: letters home, 1912 – 24 vol. 3 May 12, 1915, My dear Father.

[27] RH, MSS. Afr. S. 393. Law (Patrick John) Letters to his parents during colonial service in Northern Rhodesia 1932 – 36. Senanga, Zambesi River, Barotseland, Sept 7th 1932, My dear Father.

ancestors of the dogs on the Isunga Tea Estate when he mentioned the effects of earthquakes (which frequently rock this part of Uganda) on dogs:

> The aftershocks continued in Fort Portal for several days. We always knew because 'Rajah', the Bhimji's black Labrador, always stood up, growling, from the stone floor, somehow sensing the vibrations before we did. Household artifacts would rattle, but nothing worse. Once the dog felt comfortable enough to lie down again, we would relax. (Baker & Polonski n.d.: 83)

They were hardly the sort of dogs to intimidate and enforce 'correct' behaviour amongst the colonial subjects of the District Commissioner of Kibale (Shutt 2002: 273).[28] Kibale's current dogs accompany tourists as they go for walks or hikes around the guest house and do not bark at strangers along the way. They have developed a disposition that makes them pleasing to be with for tourists and non-threatening for the local children, workers and travellers, and they are admirably adapted for the socio-cultural setting in which they find themselves.

Tea

The tea fields of Isunga Tea Estate continue up to the edge of Kibale National Park. On the mountain slopes row upon row of bushes have been planted along the contour lines. In the valleys below and along the limits of the rainforest, undisturbed forest cover is almost permanently covered by wisps of cloud that, as the day progresses, move up the valleys to mingle with incoming clouds and rain on the fields of tea. Every morning, before sunrise, hundreds of male workers, who are summoned to a roll call by a bell, march along the contour lines and pluck fresh tea leaves within set sectors of the fields. At the side of the road at the end of these sectors, low concrete-floored open-walled shelters are used to store the baskets of freshly picked tea. In the early afternoon, trucks collect it and many of the workers and transport them to the tea factory.

Along with coffee, the production of tea has become one of the mainstays of Uganda's agricultural economy and the country's third largest foreign-exchange earner with about 90% of Uganda's total tea production being exported.[29] Uganda is ranked 13th amongst the world's largest tea producers and with over 62,000 people employed in the sector, it directly

---

28 She also describes Native Commissioner Hulley who owned a dog named Chips that would remove the headdress of any African who entered Hulley's office.

29 http://allafrica.com/stories/201109122241.html. The Uganda Tea Association estimates that the tea trade earns Uganda US$ 90 million in foreign exchange. Accessed 14 September 2011.

Photo 11.4 Tea estates and Kibale National Park [Photo: Jan-Bart Gewald]

supports about 500,000 dependents.[30] Tea was introduced into eastern Uganda close to Mount Elgon in the early 1900s, before moving westwards particularly after World War II (O'Connor 1965). The production of tea has a long and convoluted history but it was established in Uganda using British colonial experience gained in the Indian sub-continent, and Sri Lanka (Ceylon) in particular. Working relations, originally developed in Asia, were soon changed in Uganda. Nyhart's work in the late 1950s brought this to the fore when he contrasted two tea schemes being run in Uganda at the time: one by the Uganda Development Corporation and the other by the Colonial Development Corporation (Nyhart 1959). His work, now written more than 50 years ago, sounds similar to current development reports with its particular attention to 'local participation, viability, and finance and management'. However in contrast to much of the material being produced today, issues relating to gender were excluded. In itself this is interesting because, particularly on account of the fact that unlike in Sri Lanka and elsewhere in Africa (Malawi and Zimbabwe), most of the labour employed on the tea estates of western Uganda is male.[31] It

30 http://www.uce.co.ug. RIS Uganda, Agricultural Markets and Marketing Information: 'Uganda Tea Earnings Increase to $776 Million'. Accessed 14 September 2011.

31 On labour relations and labour militancy in tea production in Africa in contrast to Asia, see Palmer (1986).

has been noted that 'The greater Toro region has for long been a zone of immigration attracting migrants from as far as Tanzania, Rwanda and the Democratic Republic of Congo. Tea plantations in Kabarole are a major attraction for migrant labour from over-populated areas such as Kigezi' (Awe 2011: 60).

Given that the migrant labour is predominantly male, it comes as no surprise that this labour force has been employed on the tea estates at the expense of local labour, be it male or female.[32] Yet it does indicate a marked contrast with the way the tea plantations, modelled as they were on the Asian experience, were originally conceived in Uganda. Work by Konings (1995) on male and female workers on tea estates in Cameroon indicates that the transformation of labour relations in Uganda, in contrast to those in India, will have had long-term ramifications concerning the way tea can be produced in Uganda. The major transformation that has occurred on the tea plantations of western Uganda is that, unlike those of Sri Lanka and India, male labour is employed in western Uganda. What the impact of this is and, more importantly, why this should be so requires further research.

Lawns

A particularly striking aspect of the Chimpanzee Forest Guest House is its extensive and meticulously clipped rolling green lawns that extend up to the property's perimeter. Everyday a man endlessly traverses the lawns pushing a Ransome cylinder lawn mower. Normally all forms of vegetation, with the exception of fruit trees, are cleared from the immediate vicinity of buildings as grass can potentially house vermin and pests, snakes, rats or mosquitoes.[33] However the aesthetically pleasing nature of grass lawns and their attraction to British colonial administrators and later tourists were such that they were introduced in specific settings in colonial Uganda.

32 See comparable work by Chirwa (1993) in Malawi that indicates that Nyasaland planters could not afford a large, regular work force. The tendency was thus to rely on cheap categories of labour, predominantly women, children and migrants.

33 McCaskie (1995) has written extensively on Asante attitudes to vegetation. Admittedly his work deals with issues in a country at some distance from Uganda but it is probable that similar cultural ideas regarding forests are to be found. Comaroff (1991) has also written on the danger of untamed forests and McCann (2005) has detailed the dangers of planting maize crops to close to homesteads on account of the leaves providing excellent water reservoirs for the breeding of mosquitoes.

Photo 11.5 Lawn, ornamental shrubs, trees, hedgerow, tea estates and forest [Photo: Jan-Bart Gewald]

The development of lawns, 'piece(s) of grass kept mown and smooth in garden(s) etc.' (Allen 1985: 414), for no other purpose than aesthetic reasons is something that appears to be a peculiarly British thing that was transferred around the globe as the Empire grew. It is believed that lawns developed out of pastureland that had been extensively and closely cropped by rabbits and sheep. In the 1600s, the development of the garden in Jacobean England led to the emergence of the cropped English Lawn, which became a status symbol of the aristocracy and gentry. However, in contrast to earlier lawns, those that developed in the 1600s depended on extensive human labour armed with scythes and clippers to ensure the close-cropped nature of the grass. Formal lawns thus remained the preserve of those who could afford to employ labour for no other purpose than ornamentation. This changed in the 1830s when Edwin Budding invented the lawn mower as a superior alternative to the scythe. Budding's patent described his cylinder mower as a machine 'for the purpose of cropping or shearing the vegetable surfaces of lawns, grass-plats and pleasure grounds' that 'country gentlemen may find in using... an amusing, useful and healthy exercise'.[34] It is doubtful whether the men who mow the lawns at the Isunga Estate ever find it to be anything other than

[34] http://thelawnmower.info/history.php. Accessed 13 September 2011.

employment and far from being 'amusing, useful and healthy exercise'. Nevertheless, what is of interest here is that the consistent and continual mowing of the lawns serves no other purpose than the aesthetic, i.e. the lawns are mowed to please the eyes of tourists and to contribute to the success of the guest house. The innovation, in this instance, that has been appropriated is the concept of the aesthetic lawn.

Library and labels

One of the most influential innovations introduced into parts of Africa and Uganda was and still is literacy, namely the ability to store thoughts and concepts in a rapidly universalizing set of standardized symbols.[35] The Latin alphabet with its 26 distinct characters, which can be combined in an endless series of combinations to put forward an infinite number of ideas and concepts, is surely one of the world's greatest and most liberating innovations. Introduced into western Uganda by missionaries in the 1890s, literacy soon determined access to success within the new political dispensation that was established in what would eventually become the independent state of Uganda.[36] Literacy and numeracy are the minimum necessary prerequisites of any person seeking to make a decent living in urban Uganda today. Those living in rural Uganda, the mythical uncaptured peasantry of Goran Hayden, may be able to survive without these skills but their day-to-day lives are determined by the many and varied calculations and written words of those in power. For example, the written word determines when, who or what may enter Kibale National Park, or who may plant what crops when and where, or who pays tax and when and where. Literacy and numeracy mediate the rise and fall of the value of the Ugandan shilling, the price of commodities and, literally, the cost of living in Uganda.

Literacy may determine the difference between a life of poverty and starvation or a life of wealth and plenty. But equally relevant is the way the written word is presented. Its essential feature is to communicate ideas and concepts conceived in another time or place. Yet how the written word is presented is equally important and conveys further information than merely the concept or idea being expressed. Typesetters use

---

[35] Forms of literacy did exist in Africa prior to the 1890s and African societies did have systems of standardized symbols that conveyed concept and ideas.

[36] On the importance of literacy, see Anderson (1991: 224).

different typefaces if they wish to imbue the written text with extra meaning, to put across information that they believe adds to the concept or idea already contained in the word itself. The choice of presenting ideas in another typeface has an immediate impact on the reader that extends beyond the ideas being presented. Thus much as the graphic artist Art Spiegelman consciously adapted his script for his novel *Maus*, so too the written words displayed here and there on the grounds of the Chimpanzee Forest Guest House contain additional information.

A wide variety of distinct tree species are to be found scattered around the grounds of the Chimpanzee Forest Guest House and a number of these bear signs giving the tree's name. They are made of rough wooden planking that has been painted white and the Latin name of the tree has been written on it in hand-painted black letters. However, the words have been painstakingly copied onto the planks without any regard for the space available and it is clear that the painters had little if any understanding as to what they were transcribing. In short, the luminous idea of affixing names to plants was an attempt to answer the unvoiced queries of enthusiastic eco-tourists but the people who actually carried out the task entrusted to them had no idea as to what the planks and their labels were actually meant to communicate.

Within the Chimpanzee Forest Guest House proper there are remnants of a form of literacy that appeared to be the norm for its original inhabitant, namely C.W. Switzer. One can see that a number of doors have been affixed with a raised dark hardwood label embossed with gold lettering that has been painted in a font that recalls the height of the Empire. These marvellous labels indicate the scullery, pantry, bathroom and so forth and are undoubtedly from when the building was first constructed

Photo 11.6 Tree label on the grounds of the Chimpanzee Forest Guest House, Kibale
[Photo: Jan-Bart Gewald]

in the 1950s. Similarly, the guest house contains the remains of Switzer's personal library in glass-fronted shelves containing a number of valuable volumes of Africana relating to Uganda proper, and eastern Congo and East Africa in general. In addition, there is a fairly substantial collection of volumes of the *Uganda Notes and Records* extending up to the early 1960s, along with a variety of hardback novels published in the 1950s. This collection of books has been augmented by an eclectic collection of trash fiction and overland travellers' trade paperbacks as well as a variety of five-year-old glossy magazines in Italian, French and German.

Reading and writing lay at the heart of C.W. Switzer's life and career, as a student at Oxford, a colonial official in Uganda and a tea planter in Kibale. The reading and writing of English and later Ugandan vernaculars in the Latin script was introduced by missionaries and officials associated with the British colonial administration. Over the course of time, reading and writing have become central to the daily lives of Uganda's population, yet the ideas and concepts associated with writing are certainly no longer those of the missionaries and colonial officials. The high ideal of quiet scholarly contemplation aspired to by Switzer in his pious library adorned with an etching of Oxford has been transformed into nothing more than a jumble of printed papers that is a quaint reminder of a bygone era that is essentially meaningless to most of the visitors at the guest house. What could have been a source of knowledge has become no more than an artefact used to raise the appeal of the guest house, much as random books litter the showroom displays at IKEA and the *nouveau riche* purchase complete colour-coded libraries to emphasize their new social standing while never in fact actually reading any of the books themselves.[37] The books on the shelves in the Chimpanzee Forest Guest House were once actually read, yet now they are no more than a display and the knowledge within them is hidden.

Conclusion

The former colonial residence of what has now become the Chimpanzee Forest Guest House on the Isunga Tea Estate with its jacaranda trees, sprawling lawns, rambling bougainvillea, boisterous dogs, traces of Oxford and World War I, culinary sauces, small library, door labels and

[37] See F. Scott Fitzgerald, *The Great Gatsby* (New York: 1925). Chapter 3 describes a character nicknamed 'Owl Eyes' who pores over real but uncut and unread books in Gatsby's library.

tree names is far more than merely a magnificent place to spend time. If one chooses to look a little deeper, it provides direct insight into the way cultural and material innovations that were developed elsewhere can be introduced and transformed in a new setting. However, to truly understand these transformations in western Uganda and to extend beyond a mere exoticization of the seemingly familiar in a strange setting, it is necessary to place such innovations within their historical context, i.e. to seek to understand from where, how and why they came to be introduced and then changed in western Uganda. To do so properly, it is necessary to place western Uganda, and the Chimpanzee Forest Guest House in this case in particular, within the context not only of British Imperial and colonial history but also global history and cultural history as a whole. Only then is it possible to register the 'cultural resonances' that Gombrich referred to and to realize that what can be found on the Isunga Tea Estate are not merely the foibles of a long-dead colonial official and tea planter. Instead we get to see that the estate and all that is on it is an amalgam of influences and innovations from around the globe, from Toro to Kashmir and far beyond.

The Isunga Estate is a small mixed farm that deals in all manner of activities from guided chimpanzee walks to tea production and small-scale cattle rearing. By spreading its risks and thus ensuring its survival, the estate and Switzer's descendants draw on a world of experiences and innovations that can only be adequately contextualized and explained by reference to global and imperial history. Yet, as the example of Isunga Tea Estate shows, global and imperial history remains but an empty shell of chronological dates and names unless it is twinned with social and cultural history. Apart from dates and names, the estate is only truly understood when aspects of history of a more cultural nature are included, for example, a history of consumption, food, form and design. In other words, the full panoply of the historical discipline.

This is but a short exploration of a few themes relating to the history of transformed innovations on a farm in western Uganda, a fuller study would require a detailed investigation that would follow the history of the various innovations introduced back to their separate and distinct origins. This would not be a quick or easy exercise but would be an attempt to explore, step by step, the ancestry of innovation and the related social and cultural influences along the way. For example, tea in Uganda today is far more than merely a beverage that was originally imported from China. It speaks to the development of plantations on the Indian sub-continent and all that this entails, be it labour relations or forms of terracing. These are innovations that, following their introduction into Uganda, were transformed over time in response to local socio-cultural and/or environmental conditions.

The life of C.W. Switzer and his estate in western Uganda are an example of the way in which, as elsewhere in other former colonial settings, the former colonial overlord has become part and parcel of local society, no longer a stereotypical colonial oppressor but a successful, respected and revered ancestor. Through his life and legacy, Switzer has shown that history is far more subtle than merely violent conflict and opposition. Most of the time, people live together in peaceful coexistence, albeit on the basis of power relations that have been fought over but which inevitably change over time. As such, the life of Switzer calls for a re-examination of the late-colonial period in Africa, a time when men like him transformed themselves into new Africans.

References

Allen, R.E., ed. (1985) *The Oxford Dictionary of Current English*, Oxford: Oxford University Press.

Anderson, B. (1991) *Imagined Communities: Reflections on the Origin and Spread of Nationalism*, London: Verso.

Awe Environmental Engineers (2011) 'Updated Environmental and Social Impact Statement (ESIS) of Upgrading Kamwenge – Fort Portal Road (66.2 Km) to Bituminous Standards', Report prepared for Uganda National Roads Authority, Kampala.

Baker, R. & M. Polonski (n.d.) Cold War in a Warm Place: Memoirs of a Post-Imperialist and a Soviet Advisor in Africa. http://classwebs.spea.indiana.edu/bakerr/Cold%20War%20in%20a%20Warm%20Place.DOC. Accessed 14 September 2011.

Barfoot, C. (2004) 'Blake's England and the Restoration of Jerusalem: Too Late for Green and Pleasant Idylls?'. In: A. Gilroy, ed., *Green and Pleasant Land: English Culture and the Romantic Countryside*, Leuven: Groningen Studies in Cultural Change.

Bijker, E. (2001) 'Technology, Social Construction of', *International Encyclopedia of the Social and Behavioral Sciences*, Amsterdam: Elsevier, pp. 15522-15527.

Broadwood Johnson, T. (1912) *Tramps Round the Mountains of the Moon and Through the Back Gate of the Congo State*, London: Dana Estes & Co.

Burke, P. (2010) *What is Cultural History?*, Cambridge: Blackwell.

Chirwa, W.C. (1993) 'Child and Youth Labour on the Nyasaland Plantations, 1890-1953', *Journal of Southern African Studies* 19(4): 662-680.

Comaroff, J. (1991) *Of Revelation and Revolution: Christianity, Colonialism, and Consciousness in South Africa*, Vol. 1, Chicago: University of Chicago Press.

Crew, D.F. (1989) '*Alltagsgeschichte*: A New Social History "From Below"?', *Central European History* 22(3/4): 394-407.

de Certeau, M. (1984) *The Practice of Everyday Life*, Berkeley/Los Angeles: University of California Press.

Gombrich, E.H. (1969) *In Search of Cultural History*, London: Oxford University Press.

Hobsbawm. E. & T. Ranger, eds (1983) *The Invention of Tradition*, Cambridge: Cambridge University Press.

Hoffmann, A. (2007) 'Archaeological Investigation of Cambridge Town Allotment 389, the National (Alpha) Hotel', Report to the New Zealand Historic Places Trust and Symphony Projects Ltd.

Hummel, F. (2001) *Memories of Forestry and Travel: Uganda, Mexico, Britain, Brussels and Beyond*, London: Radcliffe Press.
Ingham, K. (1974) 'Foreign Relations of the Kingdoms of Western Uganda'. In: R. Ingham, ed., *Foreign Relations of African States*, London: Butterworth.
——. (1975) *The Kingdom of Toro in Uganda*, London: Butterworth.
James, L. (2006) *The Middle Class: A History*, London: Hachette.
Kline, R. & T.J. Pinch (1993) 'Taking the Black Box off its Wheels: The Social Construction of the American Rural Car'. In: K.H. Sørensen, ed., *The Car and its Environments: The Past, Present and Future of the Motorcar in Europe*, Luxembourg: European Commission, pp. 69-92.
—— (1996) 'Users as Agents of Technological Change: The Social Construction of the Automobile in the Rural United States', *Technology and Culture: The International Quarterly of the Society for the History of Technology* 37(4): 763-795.
Konings, P. (1995) *Gender and Class in the Tea Estates of Cameroon*, Aldershot: Avebury.
Leong-Salobir, C. (2011) *Food Culture in Colonial Asia: A Taste of Empire*, London: Routledge.
MacGregor, N. (2010) *A History of the World in 100 Objects*, London: Allen Lane.
MacKenzie, J.M. (1984) *Propaganda and Empire: The Manipulation of British Public Opinion 1880-1960*, Manchester: Manchester University Press.
——. (2011b) *European Empires and the People*, Manchester: Manchester University Press.
MacKenzie, J.M., ed. (1986) *Imperialism and Popular Culture*, Manchester: Manchester University Press.
——. (1992) *Popular Imperialism and the Military 1850-1950*, Manchester: Manchester University Press.
——. (2001a) *The Victorian Vision, Inventing New Britain*, London; V&A Publications.
McCann, J.C. (2005) *Maize and Grace: Africa's Encounter with a New World Crop, 1500-2000*, Boston: Harvard University Press.
McCaskie, T.C. (1995) *State and Society in Pre-colonial Asante*, Cambridge: Cambridge University Press.
Nyhart, J.D. (1959) 'The Uganda Development Corporation and Agriculture', *East African Economic Review* 6(2): 104-116.
O'Connor, A.M. (1965) 'The Geography of Tea and Sugar Production in Uganda: Some Comparisons and Contrasts', *East African Geographical Review* (3): 27-35.
Oudshoorn, N.E.J. & T.J. Pinch, eds (2003) *How Users Matter. The Co-construction of Users and Technology*, Cambridge, MA: MIT Press.
Palmer, R.H. (1986) 'Working Conditions and Worker Responses on Nyasaland Tea Estates, 1930-1953', *The Journal of African History* 27(1): 105-126.
Porter, B. (2004) *The Absent Minded Imperialists: What the British Really Thought about Empire*, Oxford: Oxford University Press.
Princess Elizabeth of Toro (1983) *African Princess: The Story of Princess Elizabeth of Toro*, London: Hamilton.
Rhenisch, H. (2006) *The Wolves at Evelyn: Journeys through a Dark Century*, Vancouver: Brindle & Glass.
Richards, J. & J.M. MacKenzie (1986) *The Railway Station: A Social History*, Oxford: Oxford University Press.
Ritvo, H. (1986) 'Pride and Pedigree: The Evolution of the Victorian Dog Fancy', *Victorian Studies* 29(2): 227-253.
Ross, R. (1990) 'The Top-Hat in South African History: The Changing Significance of an Article of Material Culture', *Social Dynamics* 16: 90-100.
——. (1999) *Status and Respectability at the Cape of Good Hope: A Tragedy of Manners*, Cambridge: Cambridge University Press.
Shutt, A.K. (2002) 'The Settlers' Cattle Complex: The Etiquette of Culling Cattle in Colonial Zimbabwe, 1938', *The Journal of African History* 43: 263-286.

Smith Brown, E. (1964) *Return to Laughter: An Anthropological Novel*, New York: The Natural History Library.
Swart, S. (2008) 'Dogs and Dogma: A Discussion of the Socio-Political Construction of Southern African Dog "Breeds" as a Window onto Social History'. In: L. van Sittert & S. Swart, eds, *Canis Africanis: A Dog History of Southern Africa*, Leiden: Brill, pp. 267-287.
Switzer, C.W. (1937) 'The Story of Lwabinumi', *The Uganda Journal* IV(3): 262-265.
van Sittert, L. & S. Swart, eds (2008) *Canis Africanis: A Dog History of Southern Africa*, Leiden: Brill.

12

Engine of change: A social history of the car-mechanics sector in the Horn of Africa

Stefano Bellucci & Massimo Zaccaria

The car is often seen as a symbol of Western technology in Africa, with the concomitant technology, knowledge and skills having been imported from Europe. It is argued here though that Africans did not passively incorporate the motor car into their lives but actively appropriated its use and made this technology their own. The car-mechanics sector in the Horn of Africa was dominated in the twentieth century by Eritrean motor mechanics and drivers, but this was not by chance. Cars and trucks came with the advent of Italian colonization but the peculiarities of the Italian colonial system, characterized by the presence of a large number of artisans, tradesmen and small-scale businessmen, facilitated the transfer of technical know-how. Through this close contact with Italian technical know-ledge, a class of small-scale local artisans and businessmen was born. These increasingly specialized Eritreans played an important role in the region's automobile and transport sectors, displaying an ingenious capacity to adapt European technology to fit the local context, particularly by finding innovative approaches to repairs, modifications and the production of spare and replacement parts. In this case, technology was not only absorbed by the local 'colonized' workforce but was also transformed and adapted to meet local African conditions, thereby creating new technological expertise.

Introduction

The history of transfer and transformation of technology in Africa allows us to move away from the simplistic view of a contraposition between 'tradition' (indigenous) and 'modernity' (endogenous). Technology, which in some cases was introduced in Africa by Europeans, has inevitably been absorbed, transformed and adapted to meet local conditions. Development or progress has been the result of the mixture of transfer and transformation. The general assumption in this chapter is that European technology produced transformations during the colonial period. This transfer was grafted onto a social environment that absorbed external influxes and metabolized them with local heritages. Ultimately, the process resulted in the transformation of the new technology, in this case the motor industry, in the Horn of Africa, with new mechanical technologies and social classes being created.

The automobile sector provides a particularly relevant illustration of this type of dynamic. Undoubtedly cars and trucks represent a type of technology that was developed outside of Africa and reached the continent with the advent of colonialism.[1] However, in terms of the labour market (made up of mechanics, body repairmen, car electricians, carburettor, radiator and tyre repairers, upholsterers and drivers), this technology became locally assimilated into the world of car mechanics in a short space of time. The particular conditions in which automobiles needed to operate in Africa led to considerable innovative techniques needing to be perfected through frequent modifications and adaptations, including the manufacturing of spare parts.

This innovation had a noticeable social impact that went beyond the mere provision of service. Indeed there were groups that, on the basis of this profession, established a regional social mobility that is still apparent today.

Mechanical skills in the Horn of Africa are commonly associated with Eritrean expertise and in many areas Eritrea's presence in the sector of road transport remains strong. Though less common than in the past, one can still see old Fiat 682 trucks being driven along dusty roads by Eritrean motorists. In hundreds of garages in the Horn of Africa the 'Eritrean connection' is readily apparent in the apprenticeship of the mechanics, the ownership of the business or the technical words used for the mechanical tasks that still reflect the Italian influence. The way in which this region can be viewed as a vast market with a specialized workforce in the field of car mechanics reveals the extent to which know-how was successfully transferred and to which the local mechanics were able to adapt their

[1] On the history of motor vehicles in Sub-Saharan Africa, see Gewald *et al.* (2009).

skill base to fit regional realities though their technical competence and managerial abilities.

This chapter aims to set out the historical parameters underpinning this process, particularly the way in which Eritrean artisans initially mastered the technical aspects of the mechanics involved and how they made use of this expertise to create economic opportunities on a regional scale. This historical time-line begins with the arrival of the Italians in the Horn when one of the main focuses of their colonial policy was on the construction of roads and the automobile industry as part of their ideological ambition of creating an empire. As Podestà (2002: 124-125) has shown, the model of Fascist colonialism sought to bring to Africa all of the productive elements of the mother country, namely a labour force made up of peasants, factory workers, artisans, office employees, tradesmen and small-scale businessmen, thereby rejecting the more traditional or liberal capitalistic colonization model that was organized for the benefit of a restricted white élite, favouring an increased Italian presence. Between 1936 and 1939 about 200,000 workers thus arrived in Italian East Africa, among them nearly 3,500 truck drivers.

The Italian Fascist imperial regime harboured no doubts about the central role labour exploitation should play in its expansionist programme. Mussolini himself did not fail to recall this fact on the day he made his *Proclamazione dell'Impero* to the Italian people informing them of the birth of the Italian Empire following the conquest of Ethiopia: 'The Italian people have spilled their own blood to create an empire. They will fecundate this empire with the fruits of their labour and defend it against all with their arms' (Mussolini 1936: 10).

According to Fascist ideology, Italy was a *civiltà del lavoro* (a civilization built on labour) and it therefore followed that the Italian Empire could only be an *Impero del lavoro*[2] (an Empire built on labour). Italians were described as a population of workers in search of a 'living space' to give them the opportunity to give free rein to their creativity and to 'fecundate' those countries on the margins of the 'civilized world'.

In the immediate post-WWII period, labour provided one of the central arguments underpinning Italian claims over their former colonies by the pro-American, pro-capitalist Christian Democratic Party. The fact that

---

[2] *Impero del lavoro* was the title of the fortnightly journal of the Fascist Labour Inspectorate in Italian East Africa, which was published in Addis Ababa from 1937 to 1940. Cf. Fossa (1938).

Fascism was finished did not translate into a critical evaluation of Italy's actions in Africa by the regime's successors. Indeed the new Republican government took steps to promote an improbable return by Italians to former colonies in its initial years in power in the late 1940s, which only met with any success in the case of Somalia. In Eritrea, where there had been more palpable technical achievements during the period of Italian colonialism, attempts were made to legitimize a return to Italian control over the country, emphasizing the various Italian development initiatives that had borne fruit,[3] and conversely forgetting the injustices and crimes committed by the very same colonizers.[4]

At the heart of these Italian claims was the idea that Italian colonialism was unique in the sense that, as it was often recalled, instead of seeking to exploit the colonized lands, it sought to bring about development by investing around Lira 35 billion in its colonies (del Boca 1992: 196). This is, of course, an arbitrary look at colonial history from an Italian perspective, but a hard one to overlook, even today.

The theme of labour, be it in the Fascist or the Republican period, thus remained centre stage and it was only when the history of Italian colonialism started to be appraised with greater critical attention by left-leaning and Marxist historians in the 1980s that this theme began to be viewed in a questionable light. Labour – and the construction of roads, bridges, factories and buildings – had been associated with the myth of the 'morally good, hardworking Italian' (*italiani brava gente*) and the need to escape from this rhetoric led to historians putting to one side any focus on the role played by labour in the relationship between Italy and Africa.

Our position is that labour is a key factor to understanding and evaluating the impact that the Italian presence had in Africa and that labour provides an important perspective through which to examine the processes of innovation, the way these were harnessed and the contribution of the African population *in situ* in finding local adaptations for technical know-how. This is a delicate subject because it deals with the somehow 'positive' aspect of colonialism, namely the fact that colonialism brought technological advancement to the Horn of Africa. We propose that colonialism should be looked at in dialectical terms: it was a force for change

[3] The literature on this aspect is extensive. Some authors stand out in this regard, for example, Fiore (1948a: 5) who, in addition to his monographs on the subject, wrote numerous articles along the same lines: 'For Italy not to be given back its former colonies would be an act of injustice: historically, politically and socially'. See also Fiore (1948b).

[4] On Italian crimes committed in Africa, see del Boca (2005).

in society but the changes it wrought came about in a context of domination and coercion.

The arrival of the motor vehicle and the car-mechanics trade

The first automobile made its appearance in Eritrea in 1900. Given the high costs involved in owning and maintaining these early engine-propelled automobiles, their use was largely made possible by the setting up of societies/associations. In December 1919, the Eritrean Society for Automobile Transport was founded in Asmara:

> ...with the aim of using public automobiles fitted with a taximeter, and buses for service in the cities of Asmara, Massawa and their outskirts, and potentially in other localities, buses providing a weekly transport service for passengers from Asmara to Adi Caieh, Adi Ugri, Keren and vice versa; and buses for outings; also potentially, should the Administrative Council find it appropriate at a future time, automobile transport services for goods, and services for various types of social activities.[5]

The society, funded almost entirely by Italian capital, opened its own agency in Addi Caieh in 1924, at about the same time as non-Italian entrepreneurs also began to take an interest in the transport sector. There are records, also from 1924, showing the bankruptcy of the transport business managed by Omar Hassan Sais and Saleh Mohamed.[6] An Automobile Users' Society was set up in Keren but it also went out of business in 1929.[7] One of the first instances showing Eritrean involvement in these transport societies and companies dates back to 1930 when, among the 24 founding members of the Asmara *Consorzio Esercenti Autotrasporti* (Automobile Businessmen Association), there were six Eritrean entrepreneurs, five of whom were 'automobile users', i.e. they were already active in the transport sector.[8] It would appear that there were only four or five Eritreans

---

5 *Bollettino Ufficiale della Colonia Eritrea*, (hereinafter *BUCE*)a. XXIX, no. 2, 31 January 1920, p. 29.

6 *BUCE*, a. XXXIII, no. 6, 31 March 1924, p. 118. Omar Hassan Sais also managed a bicycle-hire business.

7 *BUCE*, a. XXXIX, no. 2, 31 January 1930, pp. 40-41.

8 'Basciai Bahta Ghilamicael (deceased), born in Mulazanai (Hamasien), trader/businessman, Substitute chairman of the founding Association; Edegò Lemma (deceased), born in Adi Esà (Seraè), automobile user/businessman; Gherghis Said fu Said, born in Addis Ababa, automobile user/businessman; Salih Soliman of Soliman of Soliman, born in Asmara, automobile user/businessman; Abdalla Ibrahim fu Ibrahim, born in Aiddali (Adi Caieh), automobile user/businessman; Fiteurari Aberrà Hagos fu Hagos, born in Godofelassi, automobile user/businessman; member of the Administrative Council of the founding Association', *BUCE*, a. XL, 1, 15 January 1931, pp. 25-26.

who had driving licences in the early 1920s. Among them was Fessahtsion Beyene (1890-early 1960s) who started working as a cook and then as a driver before obtaining his pilot's licence in 1928, becoming the first Eritrean to fly an aircraft.[9]

In the early 1930s, Italian colonialism's strategic aim was to vastly improve the road network which, in view of the conquest of Ethiopia and the creation of Italian East Africa (hereinafter AOI, *Africa Orientale Italiana*), was to become the main instrument for internal communication. It is estimated that the impressive road system built by the Italians in 1938 was used by a little over 16,000 motor vehicles, two-thirds of which belonged to the Eritrean (colonial) government: 2,500 bore Government of Addis Ababa number plates, 2,000 had Somali number plates and the rest were from the territories of Asmara and Galla Sidama (Fossa 1938: 433). The 'politics of the road', which were inaugurated by the Fascist regime, presented the AOI with a well-connected and modern road infrastructure.[10] Compared to the population, the number of automobiles per 1,000 inhabitants in Eritrea at this time was significantly higher than that in Italy (Evangelista 2010).[11] They created a substantial presence that had a profound impact on the character of the area, particularly Eritrea, which was the real force behind the boom and served as the main gateway to Ethiopia in the 1930s. The transformation, which emerged in Eritrea as a result of the rapid growth in motorized transport, even affected the organization of the cities, a phenomenon that reached its apex with Decamerè, which went from being a village of a few hundred inhabitants to being the nerve centre of the Italian road network. It became a city of motor vehicles populated by about 10,000 Italians, the vast majority of whom were working directly or indirectly in the road transport sector.

Decamerè – the 'Milan of Eritrea', the 'city of motorized transport', the 'garage of Eritrea' – became the logistic centre for goods arriving from Ethiopia and heading to Massawa, a trading centre on the trading routes of Addis Ababa and Gondar and a real city of transit where truck drivers would arrive, put their trucks in for maintenance, have their tanks filled and leave again. Petrol could be found in the city at the Agip petrol

[9] On Fessehatsion Beyenesee see Issaya Tesfamariam's blog: http://kemey.blogspot.com/ Accessed 7 January 2012. See also Volterra (2005: 134-139).

[10] 'Le opere stradali', in Gli Annali dell'Africa Italiana, a. II, vol. IV, 1939, La costruzione dell'impero. L'opera dell'Italia in A.O.I. dopo la conquista dell'Etiopia, Vol. II, pp. 319-363.

[11] The number of automobiles in certain cities in East Africa had reached levels then that were only seen in Italy in the 1960s.

station, and there were warehouses and an Isotta Fraschini factory. The Capronia business was set up near Decamerè and close to the airport in Gura to repair and upgrade war planes. It was a city that housed 300 workers (Fossa 1938: 235) and its heart and soul was in the numerous small and medium-sized car-repair businesses and in the transport companies that hired Italian personnel and Eritrean apprentices as their workforce.

As late as 1946 and despite the gradual departure of resident Italians in the aftermath of British occupation, the 154 garages providing repair services in Eritrea were all Italian-owned. The greatest number were in Asmara and Decamerè (35), but there were also a significant number of garages in other important cities in the country (Mendefera: 8; Massawa: 8; Keren: 5; Assab: 2).[12] However these numbers only reflect the more structured enterprises and do not include the various itinerant mechanics who travelled to clients with their tool boxes and carried out small repairs *in situ* rather than working from a garage. The data collated from these commercial statistics cannot be seen as totally accurate because they do not take into account itinerant mechanics, who were often Eritrean,[13] and those who worked without a licence.

The birth of a motor society in Eritrea

To better understand the central importance that automobiles started to have in Eritrean society, it is helpful to consider, in addition to the services which were strictly related to the car industry, the unique significance of the recreational, social and even symbolic role of the car. From 1938 onwards, motor car and motorbike races were organized in Asmara and immediately enjoyed widespread popularity among the Italian and Eritrean public. For the Italians, the glorification of the motorized vehicle soon took on aspects related to a sense of identity. And in 1938, a crowd estimated to be some 30,000 strong attended the first Christmas Cup where cars raced around a street circuit.

The 'new Italians', i.e. those that had arrived after the conquest of Ethiopia, sought to express their own sense of identity through these races:

> It cannot be denied that 80% of us, who have arrived in this colony, share a greater spirit of initiative compared to those Italians who remained at home,

12 Data from Guida Commerciale dell'Eritrea (1946).

13 Personal communication, Mrs Maria, Asmara, 5 July 2011. Her uncle worked as a mechanic at Keren where he received his training at Serafino Giovatti's garage and then travelled from client to client, never setting up a garage.

> we have a better idea of how to create the future of our Fatherland, we have a greater spirit of adventure. And it is logical that among people of this calibre, the motorized vehicle has become a familiar object to us, one that is almost necessary to our way of living. This passion deserved to be recognized with the appropriate reward. This love of speed, this disregard for the perils involved, this frenetic and joyous sense of motion, needed to find an outlet in the most appropriate form: by means of a race![14]

The motor car thus became memorialized in local popular culture as a symbol of modernity: '[the cars] raced by, making a great racket because they had open exhaust pipes; passers-by jumped in fright at the roaring sound which was not heard every day'.[15] From 1951 onwards, the Eritrean Motor Club began organizing motorbike race meetings with a view to combining sport and leisure. On the Santo Stefano holiday (Boxing Day) in 1951 as the British were leaving Eritrea for the Ethiopian Kingdom, the first of these events was organized on a circular route (Asmara-Nefasit-Decamerè-Asmara), with an afternoon dance to follow before returning to Asmara. The Eritrean Motor Club aimed to organize weekly events with a view to 'visiting these beautiful and enticing localities in loud and cheerful company'.[16] Accepting that the competitive racing element of these events would be fairly limited in scope, the social side was emphasized, with dances, picnics and excursions. These occasions signalled a new way of interacting with motorized vehicles and one that went beyond the use of cars as only a means of transport, highlighting their value as vehicles of pleasure that could be linked to leisure activities.

In the eyes of Eritreans, the union between cars and Italians defined their perception of Italians: 'You never saw an Italian outside of a car'.[17] By the end of the 1950s, there were some 10,000 Italians in Eritrea, 8,100 of whom lived in Asmara. Of those of working age, 347 were registered as mechanics and lathe hands, while 329 were truck drivers. Only businessmen (1,750) and their staff (578) were more numerous, demonstrating the extent to which the car-related sector and the Italian community were entwined (Italian Consulate-General 1958: 49). It is important to underscore how the abolition of the Fascist racial laws in 1941 and the establishment of a local middle class created the necessary conditions for more

14 'La sagra del motore in Eritrea. La "Coppa di Natale" è entrata nella fase organizzativa', *Corriere Eritreo Sportivo*, 26 September 1938, p. 5.
15 Interview with Giovanni Mazzola, Asmara, 9 July 2011.
16 'Raduno motociclistico per Santo Stefano', *Giornale dell'Eritrea*, 21 December 1951, p. 4.
17 Interview with Tesfai, Asmara, 6 July 2011.

active participation by Eritreans in the cult of the car and other motorized vehicles. This can be illustrated by examples such as the case of Solomon Mashio, the first Eritrean motorcyclist to race competitively[18] and who enjoyed success in mixed competitions in the early 1950s. The field of car mechanics, which was so overwhelmingly Italian, is not perhaps the best place from which to observe the racial dynamics that existed in Eritrean society, although it is worth recalling that, until 1941, racially mixed car racing was banned, which demonstrated the Fascist segregationist creed. Solomon Mashio's fan base included a good many Italians, just as the clientele of the most-respected motorbike garage in Asmara were Italian. But some of Solomon Mashio's relatives recall that there was racial tension at some of his races, such as the chronometric Nefasit-Asmara race in 1954, with competitors leaving the starting line at five-minute intervals. The electrician Gianni Fusinato left the starting line in his MV Augusta 250 before Solomon Mashio but Solomon quickly caught up with him and overtook him. Gianni Fusinato abandoned the race once he saw he had been overtaken.[19]

It was probably in 1954 that an Eritrean driver participated in an official competition for the first time. In June that year, the racer G. Gheresghier came fourth in the Shiketti-Saladaro race on an Alfa Romeo 1750. A few weeks later, he took part in the Nefasit-Asmara race and on 29 August 1954 Embaiè Habtè came joint seventh in Keren while Ibrahim M. Hussayn came joint thirteenth. And again in Keren on 11 October 1953 alongside the day's main attraction, namely the Miss Keren beauty pageant, a motor-vehicle gymkhana was organized and was won by Maharì Ashera driving a Jeep.

As noted above, the motorcycle came to be regarded as a status symbol and a precious leisure tool. Italian pleasure-seekers who gathered at motoring meets started to be joined by Eritrean groups seeking a social outlet. Gradually, the motor-vehicle culture embraced Eritreans too, not exclusively as end-users, i.e. as purchasers of cars, but in all of the offshoots of the motor-vehicle sector, as chauffeurs, mechanics, car

18 The recently deceased Solomon Mashio (1926-2010) was the first Eritrean motorbike rider. He received his training as a mechanic at the garage owner by Mario Manzini, who was one of the most famous motorbike racers in Eritrea. In the early 1950s and with the assistance of a group of Italians, he became one of the first Eritreans to obtain a licence to practise as a mechanic to repair motorized vehicles. He stopped racing competitively in 1958 but remained a much-valued mechanic in Asmara. Interview with Isaa Mashio, Asmara, 22 July 2011.

19 Interview with Isaa Mashio, Asmara, 22 July 2011.

electricians, radiator experts, car-body designers, tyre repairers, dealers and upholsterers. From the 1950s onwards, there was a visible presence of Eritrean businessmen who were consolidating their professional standing and reputation step-by-step in all these professions. It is clear that Eritreans had started to master the technological know-how related to cars and to appreciate their symbolic importance by the mid-1950s.

Technological transfer through labour practice

From the Fascist period onwards, the model of Italian colonization changed radically. Africa was no longer the destination of a handful of colonials and entrepreneurs who were ready and willing to start a new life in a far-flung country. Instead, Italian East Africa came to be considered an extension of Italian society. The demographic growth in Italy was used for rhetorical purposes as justification of Italy's right to colonial expansion. During the Fascist period, there was a substantial movement of sections of the Italian population to Africa, assuming proportions akin to a *bona fide* exodus, with around 500,000 Italians taking part in the Fascist conquest of East Africa. The majority were soldiers.

Alongside workers with more obvious experience or skills that would be of use in a colonial context, such as dealers, suppliers, administrators and the military, one specific sector of Italian society moved *en masse*. This was the artisan class, which was lured to Eritrea by the state's economic incentives and which traditionally provided for the needs of the Italian community. Small and medium-sized artisanal businesses became an essential part of this new economic landscape, mainly driven by the military presence but also by the fact that the Fascists were determined to create an empire in the shortest time possible.

Small, often family-run, businesses became heavily involved in the provision of services in widely differing sectors linked to Italian colonial expansion, from transport to tailoring and from food retailers to purveyors of luxury items. The one similarity lay in the relatively small size of these businesses. They were not the type of industrial plants where the Italian management would be for all intents and purposes separate from the local workforce, with communication between the two entities being effected via local intermediaries. The usual set-up comprised a small artisanal business where the owner would be helped by a family member or an Italian associate (or partner) and a number of local apprentices. This model reproduced the Italian system and created a close working relationship between the owner and the apprentice that was well adapted for the

Photo 12.1 Solomon Mashio in his work shop in Asmara, early 1950s [Photo by courtesy of Isaa Mashio]

Photo 12.2 Eritrean bikers in promenade, early 1950s [Photo by courtesy of Isaa Mashio]

Photo 12.3 Eritrean biker posing proudly with his brand new Moto Guzzi Airone 250 cc. (Asmara, mid-1950s) [Photo by courtesy of Isaa Mashio]

transmission of expertise, technology and knowledge related to the activity in question. It is important to note that the primacy of Italians in the field of car mechanics did not imply a concomitant disregard for African labour. One should in this regard only give half credence to the well-publicized commentaries of Albert Memmi[20] on labour in the colonial context. The picture that emerges of Italian perceptions of Eritrean workers is in fact very positive: 'Eritreans are very active, extremely dynamic, well-prepared. They are keen sportsmen and good competitors'.[21] In the *bottega* (atelier, sweatshop), contact between the owner and the apprentice was characterized by harsh exploitation but also by its longevity and close confines, which in turn had an impact that went beyond the professional working relationship, for example in terms of linguistic knowledge (with locals learning Italian) and even of culinary and sartorial influence.

Even though the roles played by Italians and locals were sharply defined, with the Italians firmly in control of most of the workshops, they nonetheless worked side by side for the best part of the day. As a result of this, a pattern of unequal cooperation (but still cooperation) started to emerge that was in many respects characteristic of and peculiar to the Italian colonial experience and had profound influence on the labour market in the Horn of Africa. This situation led to the birth of a local artisanal class that at first assisted Italian businesses and subsequently replaced them, thus being recognized at a regional level as one of the most specialized and better respected workforces in the Horn of Africa. Of particular note is the fact that the Italian apprenticeship system created excellent opportunities for the transfer of expertise to the local community.

It is important to understand how this level of contact and interconnection, whilst clearly within the colonial mould, ended up in many cases alleviating the material conditions of the local workers and providing them with a profession that allowed them a certain degree of upward socioeconomic mobility. This perhaps helps to explain how the memories of the experiences of both sets of protagonists, Italian and local, are often far from negative. Some life histories have been recorded, with that of Mr Ogbai being especially interesting. He was among the first generation

20 'Nothing could better justify the colonizer's privileged position than his industry, and nothing could better justify the colonised's destitution than his indolence. The mythical portrait of the colonized therefore includes an unbelievable laziness, and that of the colonizer, a virtuous taste for action' Memmi (1965: 79).

21 Interview with Giovanni Mazzola, 9 July 2011. See also Franchini (1975-1976: 149-152, 165-175).

of local workers to gain professional training in an Italian garage and he recalled how: 'We were born and brought up with the Italians; what the Italians did, we did; what the Italians ate, we ate; like photocopies, we of the first generation share the same character as the Italians ... we have worked together'.[22]

Eritreans, who were first colonized by Italy, are aware that through their contacts with the Italian entrepreneurial class during the colonial period it was possible for those involved to obtain a transfer of knowledge that allowed them to bolster their social position and material condition *vis-à-vis* the local society. A number of these local mechanics were indeed of mixed heritage, part Eritrean and part Italian, so in a symbolic way this perhaps reflected the cultural synthesis represented by the car sector. However, the predominant reason why there was little in the way of conflict between the Italian and local labour force in racial terms was due to market-related factors. Firstly, the demand for mechanics remained constant and those with the required skills were secure in their jobs. Secondly, the types of services offered by Italian and Eritrean mechanics were not geared towards the same clientele. With Italians tending to specialize in more complex mechanical tasks and covering larger-scale orders, Eritreans filled the gap in the market by providing smaller-scale repair services for the nascent class of Eritrean car users. Even when the first Eritrean mechanics started to show interest in the area of the market traditionally covered by Italian mechanics, this coincided with the end of the Italians' productive cycle. So instead of a clash of interests, there was a gradual substitution: 'Little by little, these Eritrean assistants started to learn the trade as I had, they learnt quickly and this continued for many years until a number of Italians started to leave and the Eritrean apprentices took their place'.[23]

For many Italians, Eritrean offers to buy out their businesses constituted an excellent opportunity to obtain funds upfront for their repatriation. This was especially true at the beginning of the 1950s when an economic downturn affected both Eritrea and Italy and forced many Italians to leave, not necessarily for Italy but often for Australia or South Africa. In addition, the sale of their business to trusted local mechanics ensured that a business that they still felt emotionally involved in would survive and continue to thrive. A common theme in the interviews held

[22] Interview with M. Ogbai, Asmara, 22 July 2011.
[23] Interview with Rino Modici, Asmara, 17 July 2011.

as part of the research described in this chapter was the sense that those working in the crafting sector were deeply devoted to their profession and lived and breathed this way of life.

In terms of the exact time-line in which the transfer of technology took place, it is advisable to put to one side the chronology established on a predominantly political basis, which has up to now been used to the exclusion of all other time-lines in studies undertaken on the Italian colonial experience in the Horn of Africa. The year 1941, which marked the end of Italian domination, is of relatively less importance if one focuses on the economic picture. From an economic perspective, 1941 does not signal an end but rather projects forward, representing continuity with the part of the Italian population that remained *in situ*. The Italian exodus was a more drawn-out process that essentially drew to a close in 1974 when, with the arrival of the (Marxist military) Derg regime, the remains of the Italian community was forced to abandon large parts of their business activities and leave the country.[24] This alternative chronology provides a better-focused perspective on the interaction between Italians and locals that lasted nearly four decades.

The rising and regional expansion of a new social class

A good apprentice mechanic could learn the trade in five or six years and would then be proficient in installing and dismantling an engine and a gear box, whereas it would take around ten years to be a professional mechanic. Apprentice mechanics began their training at a very young age. Mr Ogbai, for example, began work in a garage (owned by Guido Rossi) in 1947 when he was 12 years old. Given the young starting ages and the time it took to complete an apprenticeship, most local mechanics could only hope to set up their own business when they were about 25 to 30 years of age.[25] Apart from learning the rudimentary skills of mechanics, apprentices and trainees also learnt Italian, which was the spoken medium in the workshop. Italian words are still prevalent in the lexicon of the car-mechanic sector in many languages of the Horn of Africa, especially in Amharic and Tigrinya where 90% of the technical terms used in such workshops have an Italian origin. A further indication of the close ties that have historically existed between the Eritrean mechanical sector and the first wave of Italian settlers is the use of the word *amice* to denote

[24] Eritrea, where most Italians were still located, was part of Ethiopia until 1993.
[25] Interview with Mr Ogbai, Asmara, 22 July 2011.

Eritreans born in Ethiopia. The origin of the word is linked to the Automotive Manufacturing Co. of Ethiopia (AMCE), a factory set up in 1970 by Fiat-Iveco and the Ethiopian government, which is still in the business of assembling trucks. Out of the initial workforce of 150 workers, there was a noticeable Eritrean presence, which led to the *amice* play on words.

Thorny questions revolved around how Eritreans could obtain the necessary capital to set up on their own. The usual solution was to adopt a gradual approach. Local apprentices would take on a second job and start to build up their own client base and would then use their savings to buy a piece of land and build a basic structure. The most delicate phase was how to obtain this start-up capital as once that had been achieved, local apprentices could normally obtain a bank loan in exchange for a mortgage on their property. By starting out small and buying modest pieces of land, many Eritrean artisans were able to obtain loans that in turn allowed them to take the step from being a salaried worker to being self-employed.

Even in the 1950s, trucks from Sudan were often sent to Eritrea for repair due to a lack of skilled mechanics in Sudan.[26] Saudi Arabia had also started to become an extremely interesting market, where the monetary wealth of the kingdom compensated for the lack of local skilled mechanics. The Saudi market started to take off just as Eritrea was going through a difficult economic period at the end of the 1940s when many Eritreans were leaving the country. Due to its geographic proximity and the ease with which jobs could be found, Saudi Arabia was the chosen destination for many artisans and workers from the Horn of Africa in search of a better life. The Italo-Eritrean mechanic Rino Modici remembers how, for example, in the early 1950s, one of his first jobs in Saudi Arabia was to organize a *prova d'arte* (proficiency test) for around 200 Somali mechanics who had just arrived there. This was further demonstration of the relationship between Italian colonialism and the diffusion of mechanical know-how in the Horn of Africa. And if further proof were needed that Eritrea had transformed itself into the car-mechanic centre of the region, one only needed to look at the arrival of fleets of trucks from Ethiopia to be repaired every year during the rainy season when most roads were no longer passable. They arrived in such numbers that they flooded all the repair garages in Asmara with work.[27]

---

26 Interview with Rino Modici, Asmara, 17 July 2011.
27 Interview with Rino Modici, Asmara, 17 July 2011.

In the 1960s the economic situation improved noticeably. The Emperor wanted to significantly accelerate the process of modernization in the country by ambitiously developing the nation's infrastructure. There was no doubt that the transportation network was a key element in these attempts at regeneration and, in this context, the skills acquired by Eritreans in the field of car mechanics and in certain other sectors were key.

When the local banking system started to develop in the 1960s, small car-mechanic businesses began to flourish too. There was a switch at this time from Italian to Eritrean ownership of these businesses as a result of the gradual exodus of Italians from Eritrea. Importers of vehicles and spare parts were willing to supply the lucrative Eritrean demand, fuelled by the above-mentioned credits.[28] Eritrean mechanics increasingly moved around the region in search of work and opportunities, and the intensity of the motorization process in Ethiopia was such that, in a short period of time, the demand for maintenance and repair work increased dramatically. However it also became increasingly impractical to send vehicles to Eritrea for repair[29] and many Eritrean garage businesses relocated to Ethiopia or opened a second garage there, which started to be an interesting market for Eritrean areas of expertise. Demand was high, especially for mechanics, and many firms and garages were encouraged to relocate to Addis Ababa. Substantial numbers of Eritrean artisans therefore settled in the Ethiopian capital, placing their professional services at the disposal of the country's process of modernization. Many Eritreans also relocated to Sudan and Djibouti, some moving with Italian companies.[30]

> Italians are in all parts of the world, but Eritreans are not far behind. Here we had the Fiat Tagliero, which had a number of Eritrean mechanics, who subsequently all opened up their own repair garages – some in Addis Ababa, some in Uganda, Kenya and Tanganyika, and then they expanded their businesses, buying trucks as well and providing auto transport. These are people with business nous . . . people who established auto transport businesses and repair garages . . . who learnt their trade here and then went to Germany and from there brought back machinery for the land movement, and then hired out this machinery, truly people with business acumen.[31]

28 Interview with Mauro Raffetto, Asmara, 17 July 2011.
29 Interview with Rino Modici, Asmara, 17 July 2011.
30 At the end of the 1980s, for example, the IVECO garage in Khartoum employed predominantly Eritrean personnel. Personal recollection, Massimo Zaccaria.
31 Interview with Rino Modici, Asmara, 17 July 2011.

Also of note was the Eritrean presence in the world of truck transportation, a trend dating back to the mid-1930s when the Italian Empire was focusing its energies on the creation of a road network and many Italian mechanics and garage owners were closely assisted by local apprentices, who later took over the sector when the colonial forces departed.

Innovation as transformation and adaptation

As noted above, Eritrean mechanics had to be imaginative in finding repair solutions for cars because of their limited access to spare or replacement parts and due to local environmental factors. For example, specific modifications were required because of the local terrain (Evangelista 2010) and Eritrean weather conditions also created different wear and tear on the vehicles from that experienced in Europe.[32] However the real problem lay in the scarcity of spare parts. To send car parts from Italy was expensive and time-consuming and so local garages soon specialized in the production of spare parts. The work was complicated and difficult but the lack of spare parts made it all but compulsory. Soon the level of expertise was such that it was possible to find locally made parts for the most common cars in Eritrea, which thus reduced the country's dependence on Italy. This was initially the case for the heavy FIAT 634 N (1931-1939) and for the FIAT 682 (1952-1970),[33] a medium-sized truck that can still be seen in some numbers on the roads in the Horn of Africa today.[34]

This technical expertise was further reinforced as a result of the political turmoil that engulfed the region in the course of its history. When Italy entered World War ll as an ally of Germany, the East African territories became even more isolated and self-sufficiency was the only option for them for the next five years. Indeed the years from 1940 to 1946 are still referred to locally as the 'period of isolation' (Infante 1947: 541). The situation only started to return to some sense of normality a few years after the end of the war but difficulties in finding spare parts for cars remained,

32 Another element was Eritrea's altitude. Asmara is 2400 m above sea level and diesel motors need specific maintenance at such altitude. Interview with Dante Vaccari, 12 July 2011.

33 In 1970 the FIAT 682 in its N4/682 N3 Export 'King of Africa' version was built in Nigeria. Production continued until 1990.

34 While visiting Mr Ogbai's garage, we had the opportunity to see an N3 engine that had just been repaired and to listen to his recollections about the quality of the N3 version of the FIAT 682. 'These cars can cope with difficult loads, bad roads, they still keep going . . . they never break down as long as you do a bit of maintenance.' Interview with Mr Ogbai, 22 July 2011.

propagating a continued reliance on the expertise of local mechanics. Throughout the Derg regime, it was hard to find spare parts and this situation had changed little by 2000.

Eritrean artisans have learned how to copy and modify car parts and to this day they are known professionally for this. Their capacity to recreate engine parts obviously allows for greater longevity for vehicles. Some of the parts salvaged and reconstituted, such as tyres and wheels, are fairly common, however the expertise of Eritrean mechanics also encompasses other more unusual aspects. For example, in his garage in Asmara, Dante Vaccari specializes in the repair of shock absorbers using a complicated technique. Shock absorbers that cannot be recharged need to be cut apart and the upper section replaced with a piece specifically made to measure in order to allow it to be periodically opened and recharged. The springs of the shock absorber tend to break as a result of wear and tear and, in the absence of a proper replacement spring, a used spring has to be adapted, often by shortening and tightening it. Generally speaking, the tools needed for these mechanical procedures are simple but mechanics then have to fashion a series of finely tuned tools to supplement their basic instruments for more intricate repairs.[35]

With limited access to the external market, this repair work has to be carried out using materials that can be found locally. A mechanic in Eritrea must have a deep understanding of each part he is dealing with so he can find alternative solutions and adaptations in an environment of scarcity. The difficult conditions translate into a constant push towards experimentation, which represents one of the elements of innovation in this sector. Given these circumstances, the process of repairing cars takes on a life of its own. A car is no longer a means of transport that, through regular maintenance, keeps its original appearance and characteristics. Instead it becomes the composite of its individual parts that are in a process of continual change and adaptation depending on the local availability of materials. A fairly superficial review of older cars clearly shows the level of changes made to the vehicles through this *modus operandi* in order to extend their roadworthiness. On a visit to an Iveco-licensed garage in Asmara, work was being carried out on a bus that dated from the 1980s and whose original DAF engine had been replaced by a modified FIAT 619 engine.[36] It is only natural that in such an environment, the abil-

35 Interview with Dante Vaccari, Asmara, 12 July 2011.

36 Interview with Mr Ogbai, Asmara, 22 July 2011. 'The DAF costs a ton of money and there aren't the spare parts.'

ity to adapt, to be inventive and to have an eye for detail are true assets that are much in demand. A good mechanic, according to Ogbai, 'must be hard-working, be interested in his job, (and) must use his brain'.[37]

Conclusion

To understand the history of technology in Africa, as well as the history of the African appropriation and transformation of technology, the Eritrean case, both during and after Italian colonialism, is instructive. The assumption in this chapter is that the technology under consideration, i.e. motor vehicles and mechanics, represented innovations brought to Africa by Italian colonizers through the labour sector. These were inventions that Italy itself had imported from Northern Europe at the end of the nineteenth century after the Industrial Revolution and the invention of modern motor technology half a century earlier.

In the case under consideration here, the history of appropriation and the transformation of mechanical technology reveal how Eritrean workers were capable of absorbing a new external stimulus locally. The productivity model followed by colonial Italians gave free rein to the small-scale artisan sector. And through close contact with Italian technical know-how, a class of small-scale artisans and local businessmen was born in workshops and garages. Through their labour, Eritreans were able to take the place of those who had trained them when the latter left Africa at the end of Italy's colonial experiment.

Once the transport system became crucial to the region's development policies, Eritreans made the most of their wide knowledge and this gave them an important advantage in the region and allowed them to offer their mechanical skills to neighbouring countries that were in dire need of mechanics from the 1960s to the 1980s. This regional social history had ramifications in terms of global labour history as Eritreans retained their primacy in the field of mechanics and, for several decades after World War II, were considered the experts in car mechanics in the Horn of Africa.

This occurred as a result of the need to readapt vehicles to a new environment and due to a scarcity of replacement parts. The reinvention of new technology came about through the remodelling of spare parts that needed to be constructed anew or modified to such an extent that a new and sophisticated skill base was required.

[37] Interview with Mr Ogbai, Asmara, 22 July 2011.

References

del Boca, A. (1996) 'Le colonie e il tormentato dopoguerra nell'archivio del senatore Brusasca'. In: N. Labanca, ed., *L'Africa in vetrina. Storie di musei e di esposizioni coloniali in Italia*, Paese: Pagus.

——. (2005) *Italiani brava gente?*, Vicenza: Neri Pozza.

Evangelista, G. (2010) 'L'automobilismo in Africa Orientale' http://www.ilcornodafrica.it/st-colonie.pdf. Accessed 28 September 2011.

Fiore, G. (1948a) 'Atto d'ingiustizia storica, politica e sociale', *Orizzonti Africani* III(16): 5.

——. (1948b) 'Eritrea. Brevi cenni sul suo sviluppo e sull'opera della colonizzazione italiana', *Orizzonti Africani* III(17): 7-12.

Fossa, D. (1938) *Lavoro italiano nell'Impero*, Milan: Mondadori.

Franchini, C. (1975-1976) 'La localizzazione delle industrie in Eritrea', PhD Thesis, Bologna University.

Gewald, J-B., S. Luning & K. van Walraven, eds (2009) *The Speed of Change. Motor Vehicles and People in Africa, 1890-2000*, Leiden: Brill.

Guida Commerciale dell'Eritrea (1946), Asmara.

Infante, E. (1947) *Rassegna tecnica delle industrie eritree*, Asmara.

Italian Consulate-General (1958) *Gli italiani in Eritrea nel 1958*, Asmara.

Memmi, A. (1965) *The Colonizer and the Colonized*, cited in T. Stovall (1998) 'The Color Line Behind the Lines: Racial Violence in France during the Great War', *The American Historical Review* 103(3), 737-769.

Mussolini, B. (1936) 'La proclamazione dell'impero', *Il Popolo d'Italia* XXIII(131): 10.

Podestà, G.L. (2002) 'Il lavoro in Africa Orientale Italiana (1935-1939)', in: S. Zaninelli & M. Taccolini (eds), *Il lavoro come fattore produttivo e come risorsa nella storia economica italiana*, Milan: Vita e Pensiero.

Volterra, A. (2005) *Sudditi coloniali: Ascari Eritrei 1935-1941*, Milan: Franco Angeli.

13

Water innovations among the Maasai pastoralists of Kenya: The role of outside interventions in the performance of traditional shallow wells

Moses Mwangi & Marcel Rutten

This chapter considers interventions in rural water developments among the Maasai pastoralists of Kenya and is based on personal input in this process by the lead author, who used to be a water officer in the Dutch-funded Arid and Semi-Arid Lands (ASAL) programme in the Kajiado area. This programme organized discussions in the early stages of various water projects to enable an understanding of the pastoralists' way of life, appreciate the role of local knowledge and institutions, and learn how traditional decision-making processes work. This approach has enabled an understanding of the local economy, lifestyles and the environment. A winning intercession was the improvement in traditional wells used for domestic, agricultural and livestock watering. Policy makers should try to blend the formal and the informal institutional contexts of innovations and technological knowledge applications to sustain livelihoods and ensure cooperation between the local people, government and other stakeholders. Such a fortuitous mix could become the basis of enduring engagement but innovative technological solutions will need a welcoming political economy environment for any improvements to be sustainable.

Introduction

Survival necessities have steered Kenya's Maasai pastoralists to seek innovative and improved ways of making a living through livestock keeping.

These novel solutions have gained from interaction with other communities and institutions and have improved the options available. Besides new activities that diversify household economies, efforts have been put into intensifying and upgrading livestock-holding activities. What is crucial in both diversification and intensification is secure access to water, and improvements in the design and construction of shallow wells, i.e. man-made water holes to a depth of 30 m, are vital in this respect.

This chapter considers the history of improved shallow wells, addressing both the technological and institutional aspects and showing that successful water development in semi-arid regions is possible if sufficient time and attention are devoted to building wells based on existing knowledge. However, technological innovation alone is not enough to reap the fruits of inventions as issues of political and economic power may upset the landscape and ruin progress. The development and sustainability of water resources require an understanding of the ecosystem and the role human intervention plays in modifying its properties. This chapter delves into these issues and identifies the various players that have contributed to the development of shallow wells in the region.

Transforming innovations for development

Innovation is the key to scientific and technological progress (DST 2007). Technological innovations are vital to developing a more competitive foothold in social and economic growth, and in addressing pressing developmental needs. Maasai pastoralists in Kenya have sought to strengthen their resource base by improving output and their resilience to the challenging environment they live in. With an average annual rainfall of about 500 mm and extreme weather vagaries, the availability of water is one of the most important determinants of settlement for these pastoralists. Their survival in the rangelands depends on their success in manipulating nature in an innovative fashion. When domesticated animals are used as an intermediate between grass and food for human consumption (notably milk and meat), the search is for an optimal output in this process of transformation. As a result, any pastoralist will be able to differentiate the quality of grazing areas from several angles, including the biomass quantity and quality, disease, water availability, potential wildlife predation and salt licks.

Other criteria for choosing or avoiding an area might be access to markets, schools or the risk of cattle raids. If this range of options for some innovation of natural-resource inputs and man-made conditions is less manipulative for an individual household, a more active role is required for pastoralists in the domain of innovations in the kind of livestock kept. It is not only the option of keeping several kinds of animals – cattle, goats, sheep and/or camels – that allows the owner to decide on the optimal use of the environment and whether it is best suited to grazers (cattle and sheep prefer grass) or browsers (camels and goats do well on leaves from shrubs and bushes). Adjustments in the species of cattle or small stock allow owners to innovate and improve their household economy.

In addition to striving for a specific composition of herds along the lines of sex and age, two-thirds of a herd will typically consist of female animals and a few breeding bulls while the remainder will be steers. The main issue is to have hardy and productive animals. Given the specifics of the range and the finances and labour available, pastoralists are always attempting to upgrade their traditional herds by introducing other (cross) breeds. These innovations often come at the cost of improved veterinary preventive care and medication, as well as extra feeding and watering. It is this last factor that has seen revolutionary developments in the pastoral areas of Kajiado. These improvements have been triggered first and foremost for livestock production but are increasingly also linked to processes of individualization of land ownership (Rutten 1992), sedentarization and diversification (through cultivation). Let us consider these developments in water resources, particularly shallow wells, in more detail.

The Maasai and water development: A brief history

In the mid-nineteenth century, Maasai pastoralists occupied a variety of landscapes including well-watered highlands and semi-arid grazing lands, traversing this landscape with their livestock and following the rains. They exchanged goods with their mostly agricultural neighbours and made various acquaintances. However, wars and livestock diseases weakened their once-powerful position after the arrival of colonialists in the late-nineteenth century and the Maasai were subsequently restricted to a closed reserve of lower ecological quality in southern Kenya.

Efforts have been put in place to upgrade the availability of water in this area over the years. The first initiatives were in the late 1920s when it saw the construction of water pipes, boreholes, wells and water tanks as compensation for the taking-over of former Maasai-held water sources by the Magadi Soda Ash Company, a British mining company. The driving force behind most of these water projects (sand dams, boreholes and shallow wells) was the Kajiado Local Native Council (LNC). The financial means used came from tax payments made by the Maasai themselves and from voluntary donations and labour provided by the Maasai to conduct the work required. This illustrates how keen the Maasai were to actively improve the water situation in the area.

After World War ll, the LNC remained the major driving force behind water developments in the region. The colonial era saw the introduction of deep groundwater exploitation and improved rainwater-harvesting techniques. The Lee Howl and Lister engines for water pumping have become synonymous with Maasai pastoralists' water access and management.[1] However, most water-development efforts were limited in scope and successes were minimal during this period. Boreholes in particular had low success rates and several did not produce water, broke down or dried up soon afterwards. In spite of Maasai protests, a water tax was introduced on the use of government-owned boreholes in 1950. In the same year, the Dam Construction Unit was set up to show the Maasai how to construct sand dams. Eight dams were constructed in the first year but at the end of it, only two of them contained water. Such experiences made the Maasai aware that additional efforts and other means of collecting water had to be put in place in the years ahead. The era of World Bank interventions highlighted some important issues. The initiatives in water development, which ran from the 1960s to the early 1980s, were designed to introduce comprehensive changes and encourage a move away from traditional subsistence milk-oriented production to a more commercial, meat-oriented production. The project introduced a number of water-development technologies that failed over time due to a lack of involvement by the local people.

These failures became more serious after the 1970s when reports started to appear pointing at falling water levels in the rivers in Kajiado District[2] (Map 13.1). Indeed, a comparison of the rivers in the 1930s with the current situation shows that more rivers nowadays should be labelled as 'perennial' and fewer as 'permanent'. The demand for water in the area

---

1 An example of a failed innovation is the introduction of Lamborghini engines for deep-water aquifer pumping among Maasai pastoralists by the Africa Medical and Research Foundation (AMREF). The exercise failed not only because of a lack of equipment but also because the pastoralists had more knowledge and faith in Lister engines that they had already adapted in innovative ways. Likewise, the ferro-cement water jars project funded by the Swedish International Development Agency (SIDA) through the Ministry of Health resulted in new but unused infrastructure. While the local people preferred 'mobile' tanks, the donor was adamant that the fixed tank was the correct technology. In the end, water jars littered the area as the pastoralists moved with their livestock to new locations.

2 Kajiado District was the name of the area for the specific period under review. In later years it was subdivided but re-united to form the Kajiado County since 2010.

has increased rapidly since the 1950s due to population growth, mining, irrigated (export) agriculture, deforestation and the loss of storing capacity in the rivers because of sand mining. Slowly but surely, the water-conservation functionality of the local forests and river beds is being eroded. The arrival of cultivators in the area has intensified the pressure on already-limited water resources as have the new flower, ostrich and chicken farms, private boarding schools and training institutes that have emerged in the area. All of these activities require a lot of water and careful monitoring of the existing groundwater reserves is essential.

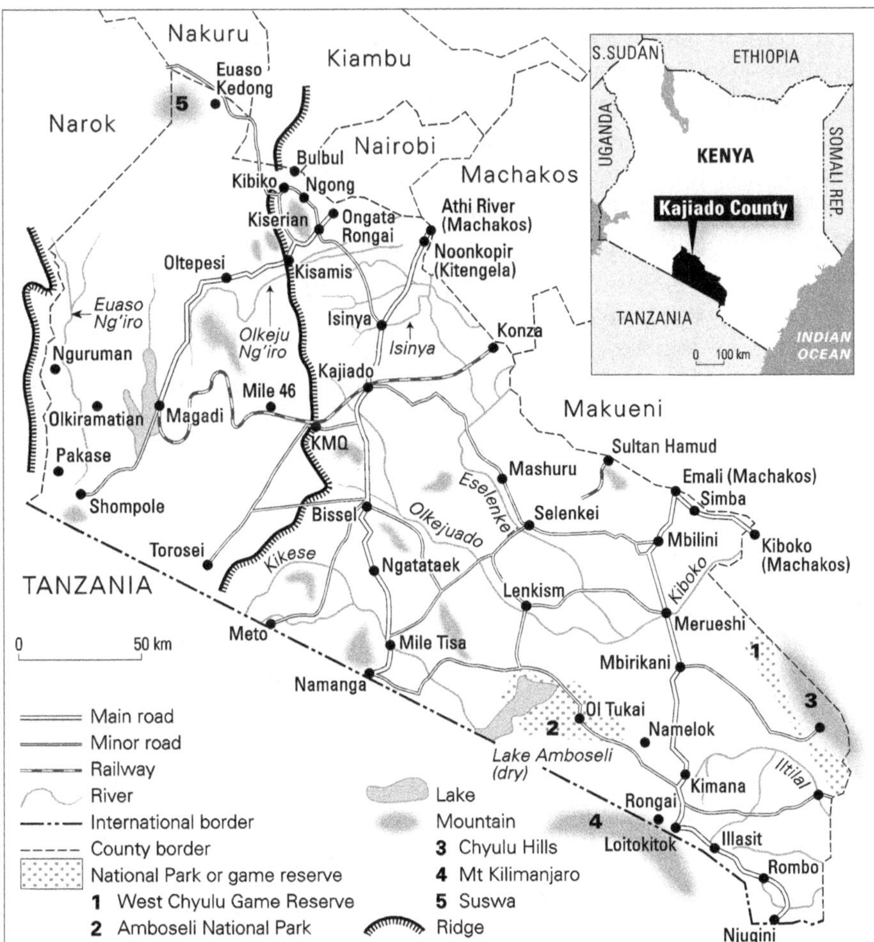

Map 13.1 Kajiado County

Another recent phenomenon is the rapid reduction in the quality of the water in the intensively cultivated regions of Kajiado. Especially in the Ngong region, which is an agricultural area that has been turned into a residential zone to the south of Nairobi, boreholes have been constructed over too small an area, which has resulted in brackish water. The use of pesticides and fertilizers for cultivation is also polluting the drinking water for both humans and livestock downstream. Mining is polluting the streams in the Kajiado area too. However, the area has also seen some success stories in water development, most notably in the development of improved shallow wells as part of the Arid and Semi-Arid Lands – Kajiado (ASAL) programme.

In 1979, the government of Kenya launched the ASAL programme in an attempt to give greater priority to marginal areas. Unable to bear the cost itself, the government sought support from Western governments and a number of bilateral agreements were signed that linked Western donors with specific sectors. In 1987, the Dutch government adopted the ASAL programme in Kajiado District to support existing institutions and 'oil the government machinery'. Its primary objective was to initiate, give direction to and sustain community action to achieve specific goals in development sectors identified by the community members themselves. This 'support' philosophy premised starting from the appreciation that a design process, as part of an intervention, aims to transform a request to implement a realistic innovation. Let us now discuss how this philosophy was executed in the field of water provision among the inhabitants of Kajiado District.

Transforming the Maasai shallow wells

Shallow wells have historically had a special place in the lives of Maasai pastoralists. Traditionally designed wells account for the bulk of the more than 5,000 shallow wells in their areas, most of which are in or along the sandy seasonal rivers.

The traditional designs faced various limitations that the new innovative interventions have attempted to solve. The newly built or reconstructed shallow wells have benefited from improved prospecting of the shallow groundwater, the construction and protection of infrastructure, and changed and better management systems.

Improved prospecting for shallow groundwater

In the traditional practices of exploring for groundwater, various plants are selected for their purposive characteristics and specific insects, animals

Photo 13.1 A traditional open well [Photos: Moses Mwangi]

Photo 13.2 Rescuing a cow from a traditional well

and rocks are used to provide assurance of the water's presence, quality and quantity. Complex human divining knowledge has been positively applied in some instances. The use of forked sticks has merged with the application of brass and steel rods, the water-bottle meniscus and the pendulum, raising confidence in the practice of exploring for groundwater in the rural areas. Association with the Mbulu people of northern Tanzania has led to an adoption of their innovative architectural exposé and knowledge of the plant kingdom for locating, developing and using shallow aquifers. However, a major limitation in the use of these local knowledge-based practices is in the certainties regarding depth, quality and volume. This has been improved with the introduction of geophysics but the application has faced cost, time and legal limitations. Combining modern knowledge with local understanding has reduced the time and financial resources needed to locate and develop groundwater resources by as much as 85%. It has also allowed the drilling of wells that can be lined and equipped with different apparatus. Maasai pastoralists' knowledge regarding locating and harvesting shallow streams has been very valuable in some remote environments. Limitations arise, however, from a lack of knowledge about how to take the practice to the next level and offer high levels of safety and sustainability. But merging indigenous innovations and scientifically based conventional wisdom has been instrumental in the installation of improved condenser and appurtenance systems.

Box 13.1 Mbulu and improvements in shallow wells in Kajiado District

The Maasai shallow wells were traditionally found along river beds where simple hand-dug holes allowed people to scope out water for livestock and human consumption. Even before the ASAL programme intervened with its shallow-well project, other groups were assisting the Maasai by suggesting improvements to their basic river-bed well design. For example, the Mbulu people of Tanzania, who were well known for their architectural prowess and groundwater knowledge, have been engaged in digging wells along seasonal rivers. Located on the banks of the rivers, instead of in the river bed itself, these wells need less maintenance, such as desilting. The Mbulu also introduced another invention, namely digging through rock, sometimes applying fire to animal manure to crack the rock. The riverside shallow wells are less likely to collapse so relocation also reduces the risks associated with the sudden flooding of rivers that can wash away both people and livestock.

The drawing of water from deep shallow wells is mostly done by throwing buckets upwards. These are then caught by a person positioned at the side of the well and passed on to another standing on a higher ridge above him/her. This creates a human ladder that allows the rapid withdrawal of water that then goes into a water trough from which animals can drink.

Improved construction and the protection of infrastructure

Traditional wells have faced major problems with silting in periods of high rainfall that have required expensive solutions. The open design is prone to pollution from erosion, silting, dust, people, wildlife and livestock. Animals and people have even lost their lives in open wells. The ASAL's water officer, this chapter's first author, and some colleagues studied several elements of the design of traditional wells, and interviews were conducted with Maasai well owners and users on this subject. This resulted in several new designs for the walls and for the mouth, the bunker, the cattle walkway and the lifting and recharging of water.

In most cases, the walls in question were lined with local stones to prevent collapse; the well mouth was reinforced to avoid problems from trampling; and the bunker's shape and walkways were redesigned to prevent siltation by trapping any material that might find its way into the well's mouth. Redesigning the well shaft and its lining meant adopting manual and mechanical water-lifting technologies. By building innovative ways to solve the problem, consultations with the local people led to tangible designs that allowed fresh perceptions and innovative engineering to be put in place. However, for an innovation to be accepted, it should be able to meet the production capacities of the old design. For example, lifting water by using a human ladder system, though highly effective with 15 litres delivery in less than a minute, was minimized to lower the chance of accidents and injuries. This had the added advantage of controlling ammonia pollution from livestock walking down the bunker tunnels. In some cases though, this resulted in a compromise. To overcome the reduced speed of water lifting, the beneficiaries wanted to have a well left open to accommodate human ladders if necessary. Another challenge was local people's fears that reducing the diameters of the well would reduce the amount of water available in it. Compromises resulted in the traditionally cut staircase being kept intact in some cases or being modified with masonry, concrete and steel to ensure safe access. With the newly introduced innovations, the dangers at the source were cut and the volume of water in the wells increased significantly. The resulting product was accepted as a combination of innovation and what the users themselves understood best and ultimately found most appropriate.

Improved management systems

Water-use traditions among Maasai pastoralists demonstrate that variations in distribution and types of watering facilities have an impact on the frequency of livestock watering. The distribution of watering points

Photos 13.3 Lining and protecting innovations on traditional wells [Photos: Moses Mwangi and Marcel Rutten]

is one of the reasons for a difference in patterns of range resource utilization. Generally, the further the pastoralist lives from water, the greater the likelihood is of watering only occurring on alternate days. Livestock owners make use of underground water, boreholes and shallow wells during the dry periods of the year but prefer to use the seasonal flows and any available pools during and shortly after the rainy season. The frequency of watering rises in the wet season when transient ponds and pools closer to home allow daily watering. By and large, the grazing management system prefers the watering of animals in the morning, with the timing depending largely on the distance between the homestead and the watering point. Drinking times are less of a constraint than the waiting time due to congestion near the water source at peak hours. Watering facilities are provided at the mouth of the shallow well for livestock and these have introduced a new order in watering.

The innovations applied to the shallow wells have resulted in the recognition of a gender dimension in the Maasai patriarchal system of water allocation. Before the introduction of shallow wells, gender issues were seldom, if ever, taken into account. For example, preference was given to the watering of animals belonging to the well's owner, followed by those of other households that had permission to share that particular well. Once all the animals had been watered, women were allowed to collect water for human consumption. Participatory gender analysis helped to highlight key gender-specific issues, such as the ownership of and access to water, water-use rights and rights to the benefits, responsibilities for maintenance and transfer rights. This has, for example, resulted in the installation of manual pumps alongside motorized ones and allows the simultaneous pumping of water, which saves women time and energy.

Photos 13.4 Gender-sensitive manual water
[Photos: Moses Mwangi]

Acceptance of appropriate water-quality standards, as defined by the World Health Organization and the Kenyan Ministry of Health, as part of the wells' management has not been very successful despite the capacity building provided. Priority by the Maasai is placed largely on the adequate provision of water, and not portability. This is one reason why the introduction of chlorine and other water-treatment methods have not replaced local practices using sand, milk, ash, the bark of acacia trees and charcoal in water-quality management. The other issues are easy access and the affordability of modern inputs. This leaves local practices to manage quality. Water sources frequented by wildlife, such as buffalo, are presumed to be uncontaminated. And innovations to make the dung on the Maasai traditional house roofs inert provide an added opportunity to harvest safe water.

Appropriation of the Land Rover 109 for rural water transportation has become synonymous with the access and management of local water supplies. In very populated areas, the vehicle is a common sight, transporting water for both livestock and people. Over time, the Maasai have developed innovative ways of operating and managing this form of transport, including recycling old tyres as shock absorbers, which reputedly last longer than the conventional ones.

Effects and impacts of the innovation

Knowing that they could sustain their traditional shallow wells with appropriate innovations, pastoralists have found themselves in a changing world of opportunities. The most transformational development has been the diversification of land use. With new water availability, pastoralists have been able to invest in dairy livestock and the growing of certain crops and vegetables. Being able to irrigate their crops with water from improved shallow wells has allowed pastoralists to considerably improve their food-security situation and expand economic practices that were previously only possible in limited parts of the region. A major advantage of this development has been that food is available for both human and livestock consumption even in times of adversity, although the need for hay has increased to support livestock in periods of stress.

Other effects of the improved access to water have been in the field of education. From a survey of 100 households it was clear that improved shallow wells had led to better school attendance rates in 52% of households. Girls, who would previously have missed classes in order to fetch water from distant wells, are now able to concentrate on their schooling.

Time has been saved and extra income earned through the sale of agricultural products and/or milk due to the improved wells near homesteads. In addition, the appropriation by pastoralist youths of technical knowledge in water development rose in 36% of the households in the first year of the intervention and enabled them to realize new employment opportunities in and outside their area of residence. Gender roles improved in all the households, with the entire family having guaranteed access to water sources as a result of improved yields and changes in resource governance. Improvements in shallow wells led to men participating in the transportation of water for household needs, generally using vehicles, 'puncture-less' donkey carts and bicycles for ferrying the water. The capabilities shown by women in managerial positions have won them the support of the men. Their acceptance in positions of leadership has been used as a platform to electable positions in society at large. Pastoralists' exposure to various social and economic institutions has led to communities being able to build linkages and networks instrumental to the welfare of society at large in areas of resource access, technology and management.

The application of new innovations has to a large extent shifted ownership, access, and operation and maintenance aspects of the shallow wells more towards individual concerns. This trend is being aggravated by land-tenure changes in the rangelands. Traditionally, ownership of water points among Maasai pastoralists depended on the type of water source. Natural ponds, for example, tended to belong to larger social units such as villages, sections or clans, while wells are more likely to be owned by an individual, a household or a group of households. The particular system often depends on the type of well, whether it is of a traditional Maasai or modern conventional type. Other factors that play a role are the amount of labour required to construct and then maintain the source. Wells that require a considerable amount of money for structural support and materials are owned by larger groupings such as clans, lineages or villages. Shallow wells dug through sand are less costly and the majority are owned by smaller groups or individuals. As a result of the subdivision of land holdings, which started in the mid-1980s, the ownership of wells is also becoming more individualistic.

Access to a well does not, however, necessarily depend on the ownership arrangement but on complex political negotiations depending on the human resources available for digging and maintenance. As long as wells are owned by large social units, the leaders can exert control over the number of wells that are constructed in any one area. If wells are owned by a family or the extended household, then there is seemingly no system for regulating their spatial distribution. It is accepted that men own the wells but not the aquifer. However, certain informal rules regulate the distance between sources, such as the need to avoid any intermingling of herds, and the fact that deep wells require considerable labour and will not be constructed unless one is a regular user of an area.

Widely respected by all groups, the traditional rules on access to shallow wells and other water resources are fundamental and seem to be taken for granted as inviolable. These fundamental rules are 'first come, first served', rights of historical precedence and rights of continued occupancy. The viability of the organization rests on a system of mutual aid, an information network and lines of power that reinforce or eliminate competition. It also defines the way resource management techniques and rules react to external forces. The rules can be described as having a 'fairness ethic', and do not require formal enforcement since they are embodied in the moral culture of all groups.[3]

Maasai pastoralists do not have a police force to regulate rules and regulations and the community relies on each member to report transgressions. Social ostracism, including social rebuke, shaming or social isolation, has traditionally been a powerful tool in keeping members in line. The belief and use of curses is another potentially powerful tool for ensuring adherence to rules. In addition, rules of reciprocal obligation are daily reinforcements of regulations concerning tenure, consumption and the protection of shallow wells and natural resources. Except for serious disputes, which come up in front of local leaders, most rules are interpreted on a daily basis by the people involved, with the goal of establishing consensus among the parties involved. There are general rules limiting access to certain pastures and water resources.

Indigenous social controls have, however, been eroding in recent years as a result of land-tenure changes, immigration, social disintegration, increasing income gaps, decreasing resource capacity and, in many cases, well-meaning development projects that have ignored traditional systems. Most importantly, the significant land-use changes, especially the expansion of horticulture and tree/flower farms, and the social and economic transformations that have taken place have increased competition for water. Zero grazing, industry and urbanization are now common and are

[3] The moral culture works quite effectively because group participants are from the same Maasai *iloshon* (sections) and clans and user groups are formed on a voluntary basis. If quarrels occur, people can either leave the group or reconciliation takes place using traditional mechanisms.

leading to increased competition for groundwater from deep boreholes. A major negative consequence of this development is the drying-up of shallow wells, a phenomenon that has never previously been seen. Nearly 100 shallow wells near flower, and to a lesser extent eucalyptus, farms in the Olkinos area alone dried up between 2005 and 2011. The seasonal shallow wells (*sinyai*) in or near seasonal river beds are also disappearing as a result of the massive sand harvesting being conducted to meet the demands of the construction industry. The area needs urgent steps to be taken to counteract the drying-up of its shallow wells. In addition to legal action and new management structures, the search is on for new technologies.[4]

Innovative leaps into the future

Shallow wells initially benefited from the problems surrounding deep boreholes, many of which were recorded as having broken down. The traditional wells, by contrast, have so far never dried up. Realizing and acknowledging this, the ASAL Kajiado programme conducted a stock-taking study to understand their qualities. Attention was also devoted to possible problems that the Maasai owners might report and this resulted in a programme to address some of them, which started in 1990 and lasted until 2005. The local economy has also improved as a result. Likewise, the sand dams and sub-surface dams, initially constructed by pastoralists using rubble and stones, have been laced with modern knowledge provided by the ASAL programme to produce structures that amplify yields from shallow wells and replenish groundwater aquifers. Other innovations include water pans, roof harvesting and rock catchments. Lately, a local NGO connected to the authors of this chapter assisted in improving another innovative water-harvesting technology whereby steam from volcanic underground sources is collected using an ingenious piping system. Improvements in local steam harvesting reinforce an earlier experience, that local innovations can be upgraded to improved and sustainable levels by external knowledge.

This chapter has attempted to demonstrate how the Maasai pastoralists have implemented their own local innovations and that these have also benefited from external developments to produce sustainable solu-

[4] The authors of this chapter are involved in this search in collaboration with the local community. A water users' association has been formed and new ways of harvesting water are being sought.

tions for their traditional shallow wells. Development of technology has nothing to do with class or social superiority but with man the producer. Most development experts, though, are yet to come to terms with the role of local innovations and science and technology in development. They are still treating innovation and technology as yet another exogenous variable. But achieving high rates of growth cannot be done without major improvements in production efficiency and output, both of which are largely dependent on the extent and rate of technological change. One view of how people come to accept new technologies, develop and innovate is that there is always a dialectic process or dialogue (Pacey 1990). The lesson from the application of new innovations to the Maasai's traditional shallow wells is that dialogue in development involves hands-on interaction with the techniques concerned, including trial and error, a coordinated response to environmental impacts and pro-active reactions to organizational and management problems. Recognition needs to be given to the fact that local people do have expertise and the expert or consultant who ignores this does so at his/her peril. Missing out on people's innovative contributions deprives local economies of the possibility of adding value to the benefit of the local people.

An important lesson for all stakeholders in technological innovations, in particular in the water sector, is that policies that are made and implemented with no regard to local situations may contribute to disasters. At the same time, disrupting local systems may have negative environmental consequences. There is thus the need to share insights with decision makers at policy or programme level in order to achieve a policy environment that is more sympathetic to and supportive of the technical contributions of the people at grassroots level. Individually, organizations may have little voice but collectively it should be possible to convince decision makers in government, as well as in multilateral and bilateral donor agencies, that many of the poorest people in the world depend on the innovatory technical skills of the local population for their survival and that their scientific and technical knowledge is too valuable to ignore.

Another lesson is that when an idea comes from outside, it may fail to be embraced by the community because it does not blend with the local reality. The process should be one that fosters joint discovery by the community and the facilitator of insights through shared experience-knowledge events that can potentially point to the solution that is most appropriate for the local reality, and perhaps different from anything previously represented within the visions of either the community or the facilitator. Development is a process where there can be no distinction between those who teach and those who learn. The sum of the knowledge base from the outside and the inside can produce a synergy that will lead to options that are not only unpredictable but that are also more appropriate than preconceived notions introduced from outside. The process of exploration and discovery is essential so that the community can grasp and internalize the information through its own experience and knowledge base.

Recognition of the value of local innovations and skills needs to be reflected in the thinking, procedures and projects of agency organizations. This may involve rethinking what is meant by the words 'technology' and 'expert', and re-examining what kind of process determines the choice of technology, what assumptions underlie the transfer of technology, at whom technical assistance is aimed, and for whom or with whom technology is designed. Importantly, any introduction of new ideas should capitalize on the relationship observed between cause and effect and not require too great a shift from the current knowledge base. Focusing on the elements that readily build on current (widespread) practices is more successful than developing approaches requiring difficult paradigm shifts.

Research on shallow wells among Maasai pastoralists confirms the need for innovations to reach out to one another and borrow from each another. This is a process based on equality and mutual respect (Mwangi 2002). To appreciate how dialogue is possible between people with conventional knowledge of innovations and others, one needs to be aware of the different knowledge characteristics of both and the strengths, weaknesses, opportunities and threats of each. While Western innovations are based on discovering basic principles that are universally valid, the strength of local innovations is nearly always related to trial, observations and experiences. But just as Western innovations may have weaknesses resulting from overconfidence in their assumptions about universal applications, local innovations have limitations too.

Conclusion

The history of water-development innovations among Maasai pastoralists in southern Kenya reveals some deliberate denial or trivialization of indigenous knowledge. Development-support agencies and researchers have in many instances portrayed local innovations as totally different and opposed to Western knowledge. The colonial era saw the introduction of deep groundwater exploitation and improved rainwater harvesting with minimal concern for local practices and rationale.

In the post-colonial era too, a number of water-development technologies failed after some time, largely due to a lack of involvement by local people. This changed when the Kajiado Arid and Semi-Arid Lands programme started in 1987. Support by this Dutch-funded programme for the sustained exploitation of shallow aquifers was revolutionary, allowing an easy fusion of local knowledge and innovations in the integrated development and management of water sources. In the ensuing technical and management re-design, perceptions of the sources were not lost, which resulted in a hybridization of hardware and software needs. This way, the Maasai have been able to absorb innovations in the water sector and ultimately make them their own. However, the current water problems faced by improved shallow-well owners, both Maasai and non-Maasai, underline the need for an institutionalized authority to safeguard these local innovations. Land-tenure changes and the subsequent diversification of land use in most parts of the Maasai region have heralded a free-for-all in the exploitation of groundwater resources that has led to what can only be described as a chaotic exploitation of local water and natural resources. Ultimately, there has been a loss of livelihoods and a destabilization of governance structures in accessing water. This acknowledges the reality that local knowledge development does not evolve in isolation and is increasingly part of other social, cultural, economic and political spheres as well.

References

DST (2007) *Innovation towards a Knowledge-Based Economy: Ten-Year Plan for South Africa (2008-2018)*, Department of Science and Technology, Republic of South Africa.

Mwangi, M.N. (2002) *The Role of Indigenous Technical Knowledge in the Development of Sustainable Rural Water Supplies*, Loughborough: Loughborough University.

Pacey, A. (1990) *Technology in World Civilization*, Oxford: Blackwell.

Rutten, M. (1992) *Selling Wealth to Buy Poverty: The Process of the Individualization of Landownership among the Maasai Pastoralists of Kajiado District, Kenya 1890-1990*', Saarbrücken/Fort Lauderdale: Verlag Breitenbach Publishers.

14

Stealing from the railways: Blacksmiths, colonialism and innovation in Northern Nigeria[1]

Shehu Tijjani Yusuf

This chapter aims to situate changes and transformations within the blacksmithing industry in the context of British colonialism. It links the massive thefts of railway keys and sleepers in Northern Nigeria to British colonial policies and highlights the technical changes and innovations that took place in the smithing industry at that time. The blacksmithing industry underwent tremendous transformations that shaped both the type of raw materials used and the technology of the craft itself. British colonial policies accelerated the decline of rural mining and smelting by starving rural smiths of raw materials. As a result, blacksmiths stole from the railways to compensate for the shortage of raw materials. Restrictions on imported manufactured metals proved a major boost to the smithing craft, leading to technical innovations and the emergence of coldsmithing. Smiths were the agents of change and innovation regardless of whether the change was triggered by external circumstances.

Introduction

Indigenous metallurgy, particularly smithing, in Africa in the twentieth century has been the focus of a number of works that have appeared over the last thirty years. One field of discussion concerns the extent to which

[1] Research for this chapter was funded by a MacArthur Doctoral Grant from Bayero University, Kano. Comments by Stefano Bellucci, Gregor Dobler and Iva Peša on an earlier version of this paper were much appreciated.

British colonial rule undermined the smithing craft, with some arguing that it caused major transitions in the industry because it led to the importation of European scrap metal. This deprived smiths of the raw materials they required and led to a decline in smelting and smithing crafts, and subsequently led to thefts from the railways to compensate for the lack of iron available (Jaggar 1973; Herbert 1993; Cline-Cole 1994; Emeagwali 1997; Aliyu *et al.* 2008). This is, however, an overly simplistic view of the impact of colonialism on the local smithing industry. According to critics, the importation of metals from Europe did not necessarily lead to an absolute decline in the local smithing craft but it was disruptive and subversive to indigenous metallurgy, especially mining and smelting. And smithing not only appeared to escape decline but even expanded as the result of a disparity in prices (Hill 1972; Hopkins 1973; Jaggar 1994).

Blacksmiths stole from the railways to compensate for the shortage in locally mined iron, which was the industry's original raw material (Fika 1978; Aliyu *et al.* 200; Yusuf 2011b).[2] But this leads to further questions about the role of theft from the railways in the expansion of smithing. Why were thefts more prevalent in rural than in urban areas? How did colonialism affect the supply and exchange of raw materials? How did the iron shortage encourage both technical and entrepreneurial innovations in the smithing industry? Who were the drivers of these innovations? This chapter attempts to answer these questions and considers the impact of British colonialism on blacksmithing in Northern Nigeria, arguing that the craft of smithing underwent major transformations in the twentieth century that accelerated the industry's decline. Stealing from the railways provided an important supply of metal for blacksmiths, especially those in the rural areas and influenced the industry in numerous and sometimes unexpected ways. The technical and entrepreneurial innovations in the smithing industry were the result of the smiths' resourcefulness and kept the industry going, despite all the odds. The smiths developed an import-substituting metal industry based on the metal stolen from the railways.

An overview of blacksmithing in Northern Nigeria

Blacksmithing is an ancient African craft dating back to the first millennium BC. While some claim that smithing was imported from elsewhere (Herbert 1993), others argue that it emerged independently in Africa (Aliyu

2 Literature dealing specifically with metal thefts from the railways is limited and the subject is often dealt with only in passing.

et al. 2008). In any case, the craft of iron working is a long-established one in Africa. Iron was smelted at Nok, in Northern Nigeria, around 500 BC and by the fourth century AD the knowledge of iron working had spread across the region. It could not have developed without prior experiences in copper and alloys and knowledge of iron working played an important role in the region's development. Some of the states were even founded by blacksmiths, suggesting a long tradition of iron working. However metallurgical implements not only placed power in the hands of builders but also in the hands of destroyers of towns and states (Herbert 1993; Aliyu *et al.* 2008; Hopkins 1973).

Over the years, blacksmiths came to play an important role in the political economy of the region. In Northern Nigeria, blacksmiths manufactured farming implements, domestic utensils and weapons, all of which were in great demand. Successful smithing depended on being able to access iron ore deposits and charcoal but iron ore differs in quality from one location to another (Herbert 1993; Aliyu *et al.* 2008). Some areas within the modern states of Katsina, Zamfara, Bauchi and Birnin Gwari had good quality iron ore, while Kano Emirate and the Bornu area had low-quality ore. The same variation in quality applied to firewood. For example, Zamfara and Bauchi were known for having good-quality fuel wood (Nalado 2004; Jaggar 1973; Herbert 1993; Aliyu *et al.* 2008). Where raw materials were not readily accessible, blacksmiths procured them from other regions, which led to long-distance trading.

In Northern Nigeria, as elsewhere, smithing was differentiated from smelting. They were distinct occupational groups requiring different techniques and knowledge. Generally, blacksmiths did not mine or smelt themselves but relied on professional miners, smelters and traders for their supplies of iron ore. Smiths who wanted to mine or smelt needed permission from their chief before they could do so. But there were some exceptions. As Herbert (1993) pointed out, the quintessential smith was one who not only forged iron but knew the secret of smelting as well. Evidently, smelting was considered part of a smith's duties among the Anka of Zamfara and Adar of Niger and smiths were expected to master the relevant skills required (Pilaszewicz 1991; Nalado 2004). In contrast, smiths in Kano, Fika and Katsina Emirates relied on intermediaries such as miners, smelters and traders for their supplies of raw materials. Most urban smiths also depended on rural miners, smelters and traders, including rural smiths (Jaggar 1973; Nalado 2004; Aliyu *et al.* 2008). The Kano smiths that Jaggar (1973: 24) interviewed claimed that their ancestors did not smelt but relied on rural networks of supply to urban areas.

Urban smiths generally procured their raw materials from rural networks (Nalado 2004; Aliyu *et al.* 2008; Ogundiran 2005; Pilaszewicz 1991; Hopkins 1973; Herbert 1993).

Like most crafts, smithing was organized around the chief of blacksmiths, whose office is hereditary. He plays an intermediary role between smiths and the traditional institutions. Smithing is a male-dominated, hereditary craft, with sons being apprenticed patrilineally or occasionally matrilineally and smiths teach their male children the craft from a young age. Due partly to Islamic influences and taboos, women have always been barred from taking an active part in smithing, except on special occasions. Nevertheless, young girls could assist in the forge. Through payment and apprenticeships, those who did not belong to a smithing family could be initiated into the craft, although there were differences between the towns. Unlike other traditional crafts, blacksmiths had certain rituals and cultural performances that set them apart from ordinary people, such as being able to play with fire. As masters of iron and fire, they were seen as cultural heroes who possessed magical powers, which differentiated them from others. The process of iron working itself possessed some magical properties too, making those who worked with iron objects figures to be revered. They were seen as intermediaries between the natural and supernatural domains. Depending on the town, blacksmiths have enjoyed positions of different status in Hausa society. For example, the Hausa smiths in Kano and Anka enjoyed a high status, whereas their counterparts in Zaria and among the Kanuri people of Bornu had quite a low status. In fact, among the Kanuri, smithing was considered a shameful occupation because blacksmiths made weapons that were used by enemies of the Prophet Mohammed (Pilaszewicz 1991; Herbert 1993; Jaggar 1973; Jaggar 1994; Nalado 2004).

By the end of the nineteenth century and as a result of both human and natural factors, smelting activities had declined, which resulted in a shortage of raw materials. Widespread deforestation of hardwood trees, the exhaustion of iron-ore deposits, changes in climate, and soil deterioration and erosion had all exacerbated the raw-materials crisis. This was further worsened by incidences of warfare, drought and famine that seriously disrupted economic activities (Goucher 1981; Pole 1982; Okafor 1993; Cline-Cole 1994; Herbert 1993; Darling 1989; Nalado 2004). However, blacksmiths kept the industry going despite the crisis, evidencing their resourcefulness and ability to innovate. As Goucher and Herbert both point out, the increased penetration of European metals into coastal parts of West Africa from the seventeenth century onwards helped to sustain the industry by providing it with substitute raw materials (Goucher 1981;

Herbert 1993). It is difficult to ascertain how far inland the metals penetrated or the extent of their impact. Certainly, the demand for imported iron reflected an existing 'iron hunger' in much of Africa, as the demand exceeded the supply from indigenous sources (Pole 1982; Herbert 1993). By the turn of the twentieth century, iron smelting had declined to such an extent that it was confined to areas close to sources of raw materials, i.e. iron-ore deposits and fuel wood (Cline-Cole 1994).

British colonialism and its impact

The advent of British rule early in the twentieth century significantly influenced the landscape of Northern Nigeria as British colonialists implemented reforms that affected the region's political and economic activities. To start with, British rule disrupted the blacksmithing craft by banning the bearing and production of weapons. The British in effect denied local smiths a source of income because they simultaneously encouraged imports of foreign weapons (Jaggar 1994; Jaggar 1973; Aliyu *et al.* 2008). Furthermore, a land proclamation was promulgated that vested ownership of land and resources in the British, which deprived smelters of access to land and minerals. The British also undermined the craft of smithing by promulgating laws that regulated the felling of trees. Such laws disrupted smithing, which depended on wood for fuel. The felling of productive economic trees for firewood was restricted, as proclamations declared some trees to be 'protected' and exploitable only by licence or permit, and a levy was also introduced on the commercial exploitation of trees. The negative influence of such proclamations can be gauged, especially as blacksmiths relied on smelters and charcoal burners for raw materials. As a result of the poor status of smelters and charcoal burners, permits were not only difficult to come by but the levying of fees discouraged smelters from felling trees and, by implication, reduced the quantity of raw materials available to smiths (Cline-Cole 1994; Aliyu *et al.* 2008). In addition, the British introduced a series of proclamations that disrupted iron-ore smelting and thus further jeopardized the smithing craft. As Jaggar, Aliyu and others attest, the enforcement of these proclamations restricted mining activities and caused the closure of smelting furnaces in Pauwa District of Katsina Emirate (Jaggar 1994; Aliyu *et al.* 2008). On top of this, the British introduced an 'iron scrap policy' that encouraged the importation of European scrap metal and accelerated the eventual decline of smelting by making the craft increasingly dependent on scrap metal (Emeagwali 1997).

The arrival of the railway in the region signalled by far the most transformative impact of colonialism in the area. The railway was central to British imperial policy, being seen as a means of 'developing' and 'opening up' the region. It was thought that without a modern transport system like the railway, a landlocked region such as Northern Nigeria could not develop economically. The railway served the political, economic, strategic and military needs of British colonialism (Lugard 1965) and its arrival in Kano in 1911 was the culmination of British imperialism in Northern Nigeria. As it opened up the region to European trade, it had an enormous impact, stimulating the agricultural export trade, particularly that of groundnuts which, from 1912 to 1914, amounted to some 21,900 tons from Northern Nigeria alone (Hogendorn 1978; Okediji 1972).

The railway promoted spatial interaction between southern and northern Nigeria. Prior to its arrival, contact between the two regions had been limited. Extensions to the rail network from 1911 through the 1920s facilitated accessibility with the outside world. This started with an extension of the line from Zaria Junction to the Bauchi tin fields, which was completed in 1915. Another line between Port Harcourt and Kaduna connected northern and eastern Nigeria and was finished in 1927. The Zaria-Funtua-Kaura Namoda line was completed in 1929 and the Kano-Nguru line followed in 1930. The last to be opened, in 1964, was the Bauchi-Maiduguri line (see Map 14.1). These extensive rail networks facilitated the emergence of new towns along the railway lines, lending them new economic and strategic importance (Yusuf 2011c; NAK 299/S.3, vol. 1 railway publications).

Effects of the railways

The railways impacted on the smithing craft due to the supply and distribution of European scrap metal that flooded the local markets. As already mentioned, European scrap metal had found its way into coastal markets before the arrival of the British.[3] Nevertheless, the arrival of the railway almost immediately increased the amount and variety of scrap metal, except in some of the more remote areas (Jaggar 1994).[4] The railway facilitated its supply and distribution across Northern Nigeria, resulting in the gradual replacement of locally mined iron as raw material for blacksmiths.

---

[3] As Pole (1981: 510) pointed out, an iron bar worth 6s 3d in Lisbon would sell for 11s 3d on the West African coast and for double this price inland.

[4] Although scrap metal was distributed across the region by rail, supplies were mainly restricted to urban railway terminals.

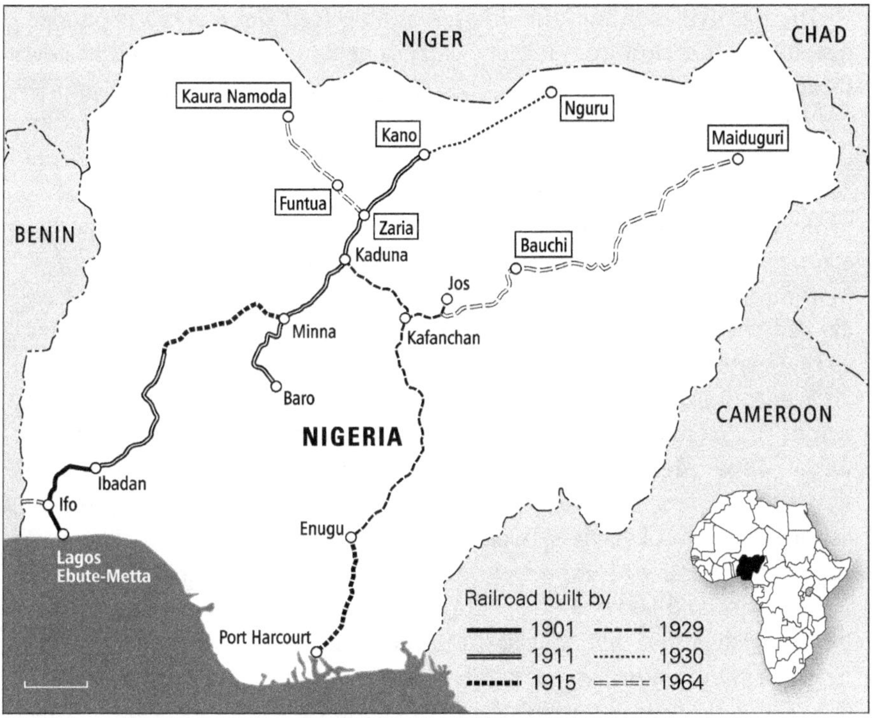

Map 14.1 Nigerian railways

The cities of Kano and Zaria became distribution points for scrap-metal supplies because of their location on the railway line, their proximity to major markets, as well as their low overheads (Jaggar 1994; Aliyu *et al.* 2008). As scrap metal became available, blacksmiths increasingly came to depend on it as a raw material, which resulted in the gradual replacement of locally mined iron by scrap metal, which was completed by the 1920s or 1930s (Jaggar 1994; Herbert 1993; Aliyu *et al.* 2008). Equally, smithing and other craft industries are said to have declined when they were faced with imported European manufactured goods (Mabogunje 1968; Callaway 1968; Shea 1974/1977).[5]

[5] Mabogunje (1968) reported a similar decline in the local craft industries of Yoruba towns; Callaway (1967) commented on the decline of smithing and other local industries in Ibadan City, and Shea (1974/1977) observed a decline in the local textile craft in Northern Nigeria in the face of competition from European manufactured textile materials.

There is evidence that the urban smithing craft thrived and expanded despite competition from imported scrap metal and it developed in new directions later in the colonial period (Jaggar 1994). However, rural smithing seems to have suffered from the scarcity of raw materials and subsequent thefts from the railways should be seen as an indication of the raw-material crisis in the smithing industry.

If the smithing craft did indeed expand, as was claimed, how then did scrap iron and other forms of imported metals fit in with the existing raw materials available to blacksmiths? Blacksmiths had used imported metals and cheap and low-quality imports, like iron roofing sheets and bars, as raw materials. Although this only occurred on a small scale, such metals were imported by the agency of Hausa smiths and traders, some of whom pioneered the scrap-metal trade. The arrival of the railways and further investment in road transport facilitated the penetration of new forms of scrap metal such as iron bars, railway sleepers and keys, as well as metal from broken river boats. As the literature argues, local iron was superior in quality to imported scrap metal. Tools made from local iron were harder and lasted longer than those from imported or scrap iron. It has been suggested that at least three hoes made from imported iron were needed to complete farm work in one season, whereas a hoe made of local iron could be used for more than two seasons. Much time would thus be devoted to the periodic maintenance of farm implements by smiths, including sharpening, straightening and reshaping tools (Pole 1982; Herbert 1993; Jaggar 1994; Aliyu *et al.* 2008).[6] Archival records show that blacksmiths derived a high proportion of their income from this work (KSHCB MLG 16490/1932 Kumbotso District Kano Emirate Reassessment, NAK MLG 9117/1929 Kura District of Kano Emirate Reassessment). Yet blacksmiths preferred imported iron to local iron for technical reasons as it was easier to repair if tools broke or became blunt, and it was easier to work since most of it was already shaped and needed little or no reshaping. This contrasted with locally mined iron, which was often too hard and brittle to repair. Its solidity made local iron particularly difficult to cut into the desired shape (Pole 1982; Jaggar 1994).

Pole suggests looking beyond the question of raw materials shortages and price disparity when discussing why blacksmiths adopted imported iron in spite of its inferior quality. Indeed how labour was organized in the

---

6 As Pole (1982: 507) points out, the low quality of imported iron caused smelters of Sukur in northeast Nigeria to start smelting in the early 1960s.

iron smelting and smithing process was important as iron production was a complex, lengthy and labour-intensive process. The economic pressures brought about by the introduction of scrap or imported metal differed, depending on whether smelting and smithing was performed by the same group or by different individuals. Smiths who smelted their own bloom could switch to buying imported iron as soon as this was advantageous. If smelting and smithing work was not performed by the same group, smelters were cut off from the market for the finished products to which their labour had contributed. This situation made them vulnerable to competition from other sources of iron (Pole 1982). This was not the case when smelting and smithing work was done by the same person and imported iron should be seen as a substitute for labour rather than as an alternative raw material. Even if smelters or smiths could afford imported iron, the proposition would not have been attractive unless there was pressure on their labour resources. These could have come from other markets with the arrival of the British. Anything that reduced production time would increase profit margins. In other words, giving up smelting and concentrating on smithing alone meant that iron workers would gain time to earn extra income. Given the inferior quality of the tools made from imported iron and their short lifespan compared to those of local iron, the use of imported iron would merely further demand (*Ibid.*).[7] Preferences for local iron also had to do with rituals attached to the material, which did not stop with the use of imported iron. As long as smithing work remains necessary and imported iron still requires reshaping by smiths, this influence is likely to continue. An additional technical reason for the preference of imported iron over local iron is that tools made of the former can easily be fired and mended when they break or become blunt, as the metal is softer than that of locally mined ones (Pole 1982; Jaggar 1994).

It is not surprising that as scrap metal made inroads inland, it gradually gained acceptance among smiths who began to use it as raw material. The penetration of scrap metal was no doubt a favourable development, especially in the light of the raw-materials shortage that plagued the smithing craft and smiths gained access to a wider variety of materials. The availability of scrap metal not only increased smiths' productive capacity, it also resulted in technological innovations. Scrap metal provided labour-saving devices at a lower real cost per unit and in less time compared to local iron. Blacksmiths turned scrap metal to their own advantage by

[7] The iron quality also determined the price and income potential of the smiths.

producing products that met local demands. As Jaggar (1994: 35) shows in his work on Kano City smiths, a shilling's worth of scrap metal could produce as many as two dozen hoe blades, each of which would have sold for 5d to 6d between 1911 and the 1920s. This would have been sufficient to buy between four and five bowls of corn and allowed the smiths to earn more than was previously possible. Pole (1982: 506) reported similar findings among Lawra smiths in northwestern Ghana where a hoe blank or flat piece of iron with a tang could be bought for one cedi. Smiths charged between twenty and thirty pesewas for converting the iron into usable tools or, alternatively, they could buy scrap iron from which they made hoes that they sold for the same price. The work would have taken the smiths about thirty minutes in the first case and about two hours in the second and in both cases, the smith would have earned about four cedis a day.

There is no doubt that, in general, the arrival and penetration of scrap metal had positive and transformative effects on smithing, at least in those areas within easy reach of the railways and major markets (Pole 1982; Jaggar 1994). Some have argued that rural smithing expanded because of the rural smiths' competitive prices (Ogundiran 2003; Hill 1972). In the case of Northern Nigeria, the opposite was true and the arrival of scrap metal may have benefitted urban smiths, whose proximity to railway terminals and major markets where the scrap metals were offloaded provided direct access to a new type of raw material. However, the distance of the rural areas from such urban terminals reduced the likelihood of the metal reaching the rural areas.

Although some of the scrap metal may still have found its way to the rural areas through the agency of local iron dealers who propelled the metal trade in the urban centres, this was erratic.[8] The confinement of scrap metal to urban markets reversed the traditional trade relations whereby urban smiths depended on rural miners, smelters and smiths as well as traders for raw materials. Mining and smelting had declined in most rural areas.[9] Previously, urban smiths relied on rural networks for the supply of raw materials. As the scrap metal made inroads inland, rural

8 Local iron dealers played an important role in popularizing iron across villages in the remote areas. Given the bulky nature of this type of metal, iron dealers could not reach all areas.

9 Sasson (cited in Pole 1982: 507) pointed out that it was only in areas where smelting was done by farmers (as in Sukur) and not by smiths that it was likely to be resurrected. According to him, such resurrection could have taken place only within fifty years at the most because later the specialized knowledge would have been forgotten.

smiths came to depend increasingly on networks of urban iron dealers for their supplies. The penetration of scrap metal accelerated the eventual decline of rural mining and smelting, forcing rural miners, smelters and smiths to abandon their craft for something entirely new. The few that held on to the trade resorted to stealing directly or indirectly from the railways to compensate for the loss of raw materials (Pole 1982; Jaggar 1994; Aliyu *et al.* 2008; Yusuf 2011a, 2011b). Smiths in areas far from urban centres were short of raw materials for much of the colonial period and this resulted in thefts from railways. Throughout the colonial period, railway sleepers were made of metal rather than wood or cement, as was the case in most of Europe at the time. The metallic nature of the sleepers and keys were thus attractive to smiths who were in dire need of raw materials. In the nineteenth century too, railways in Europe were plagued by thefts of metal.[10] In fact, even today it is relatively common to hear of the cancellation of train services in Europe due to copper and metal-related thefts. In Northern Nigeria, the railway metal thefts, which started on a small scale, have over the years turned into an epidemic that plagued not only the railway system but disturbed the British more than any other crime did.

Stealing from the track

Between 1914 and the 1950s railway metal theft assumed disturbing proportions and attracted the attention of British colonial officers in Kano, Zaria, Bauchi Plateau and Benue Provinces. The first documented evidence relating to it focuses on Dumbi and Rahama villages in Zaria Province where 35 people were convicted of the theft of railway keys in 1914 (NAK ZARPROF 2561/1914). And the annual report noted that the Emir of Zaria 'was instructed to keep a close watch on blacksmiths' huts near the line' (*Ibid.*). Although the records did not explicitly mention blacksmiths as being responsible for the thefts, their rural character and the idea of keeping smiths under surveillance have relevance for this study. The idea of keeping a watchful eye on blacksmiths may not have been mere suspicion. As raw materials were in short supply, smelting had declined and scrap metal was scarce in the rural areas. And in 1915 another group of 27 people were convicted in Zaria Province of stealing railway keys. As the report points out, the principal recipients of the keys were blacksmiths (NAK ZARPROF 2562/1915).

10 I would like to thank Mossimo Moraglio for drawing my attention to this issue.

That the document mentions smiths suggests that blacksmiths were directly or indirectly stealing from the railways. Either blacksmiths themselves stole from the railways or thefts were carried out with their knowledge and to their advantage. For instance, while commenting on the frequency of such thefts in 1921, one senior colonial official remarked that '(t)he best remedy for (the thefts) is for the Railway (authority) to cheapen the price of iron by importing and (…) distributing it to the public' (NAK SNP 17/2 11959). Even if iron was available, it was not affordable and blacksmiths could be short of iron because of its high price rather than due to an actual shortage. Furthermore, in 1945, some blacksmiths in the southern part of Zaria complained that they were not working because 'they were having no irons to work with'. This can be taken as an indication of the raw-materials shortage, which is little reported in the literature (NAK ZAR Prof RLY/2 offences on trains and railway stations). The Superintendent of Police reported in 1945 that '(s)teel is almost unobtainable these days and in consequences commands a high price (*Ibid.*). Similarly, in the 1940s, the Resident of Zaria Province also observed that '(h)oes, knives, etc. are in short supply', linking this to the spate of thefts taking place across the region. He therefore suggested that 'If unserviceable (iron scrap) could be collected and sold, it would (a) prevent it being stolen, (b) supply some of the input, which would otherwise lead to the essential need of farmers etc. being supplied by thieves' (NAK ZAR Prof 570 theft of railway materials). Iron was clearly scarce or available only at a high price, which made railway keys and sleepers vulnerable to theft. As Pfaffenberger (1992: 506) suggests, people who lose out when a new thing is introduced often devise strategies to compensate for their loss of self-esteem, social prestige and economic independence. This involves 'appropriation (…) (of the) artifact from which (they have) been excluded' and includes stealing from the railways to compensate for the shortage of iron. Some technical advantages of working with scrap metal were stability in prices, quality and the nature of the iron available to smiths. Whereas the price, quality and the physical nature of the locally mined iron were uncertain, that of scrap metal was fairly standardized and predictable.

Although metal thefts persisted, the British administration found it frustrating and difficult to apprehend the thieves. It was noted that the thefts were carried out 'first in one section of the line, then on another' (NAK ZARPROF 2475/1917). The rate at which they occurred alarmed the British, who started offering rewards in exchange for information that would lead to the arrest of the thieves. However it was unlikely that villagers who were experiencing shortages of farm implements would report the arrival of iron in their villages, even if they suspected it to have been stolen. The desire for iron and the ready-made market for stolen metals stimulated demand for stolen metal, which changed hands, moving from the immediate area of the theft to places far from the lines.[11] Stolen railway keys were obtained in Katsina (which had no rail facility at this time) at one shilling each, against a selling price of three pence in Zaria[12] (NAK ZAR PROF 2475). This suggests a decline in blacksmiths' raw materials, which stimulated a demand for stolen metals, and the further away the metals were marketed, the higher the prices they fetched. It is likely that most of the stolen metals were converted into tools or other artefacts, which explains why the perpetrators were rarely caught.[13]

The import restrictions resulting from World War I created a ready market for local substitutes, with a corresponding demand for raw materials that resulted in further thefts. For instance, 69 people were convicted of theft and receiving railway keys and sleepers in Zaria Province between 1917 and 1918 and most were arrested during a systematic search of blacksmiths' houses (NAK ZARPROF 2475, NAK ZAR PROF 163/1919). Although it is not clear whether they were all smiths, they were certainly connected to smithing in some way. Such reports evidence, furthermore, that not only smiths and/or their cohorts were involved in the thefts but that railway workers were involved too (NAK ZARPROF 2475, NAK ZAR PROF/RLY2). As a result, about forty local police were stationed at intervals along the line, especially between Kano and the Zaria border (KSHCB 318P Kano Province Report No. 57 for Half year Ending 30th June 1919). As argued elsewhere (Yusuf 2011b: 9-10), the Native Authority and the railway police were numerically unable to police the railways effectively.[14]

The increased demand for Nigerian groundnuts after the end of World War I stimulated production throughout the region and farmers required

---

11 During an investigation in Kaduna District of stolen railway metals in 1955, the police were told that the thieves had come from as far away as Abakwa, Unguwan Shanu and Tudun Wada, all in Kaduna District (NAK ZAR PROF 570 theft of railway materials). Other records show that the thieves might have come from another village because footprints leading to Doka and Kurmi villages (about ten miles north of Iri station) were discovered at the site of a theft near Kutaru station mile 529 (NAK ZAR PROF RLY/2).

12 Hill's (1978: 195) argument that smithing expanded in Katsina despite imports of metals is questionable.

13 As Anderson (1986: 399-415) pointed out, thieves forged brands on stolen cattle to avoid being detected.

14 Because of the ineffectiveness of the railway police, the native authority police were often asked to investigate cases of metal thefts but they were unable to curtail them.

more farming implements that in turn led to a higher demand for blacksmiths' products. This resulted in increased thefts of railway keys across the region, which elicited an official response aimed at curbing the menace (NAK SNP17/2 11959 Railway Ordinance 1920 on council applying).[15] Between 1921 and 1925 large numbers of railway keys were reported stolen across Northern Nigeria. In Kano Province alone, 2100 railway keys were stolen in 1924 compared to 1013 in 1923, 1030 in 1922 and 3636 in 1921[16] (KSHCB SNP/9/12/1924 Kano Provincial annual report file 635). These appear to have been the largest metal thefts from railways in the area. At the same time, the figures show the availability and scope of markets for railway metals.[17] As the annual report and other records attest, several searches and raids on the huts of blacksmiths proved ineffective. The only solution appeared to be the extensive sale of the 'scrape-iron' to local dealers (KSHCB SNP/9/12/1924; NAK SNP17/2 119.59; NAK ZAR PROF 570).[18] Interestingly, the report recognized blacksmiths as suspects or beneficiaries of the thefts. The report also acknowledged the role of local metal dealers in the iron trade, as it suggested the sale of scrap metal to them. In Zaria Province too, about 3200 keys were reported stolen in 1925, which represented a large percentage of the stolen metal in the region as a whole in that year (NAK 2572 1925 Zaria Provincial annual report for 1925). And thousands of railway keys were also stolen on the Bauchi-Rahama line (NAK SNP 17/2 119.59).[19]

The economic depression in the 1930s increased the prevalence of metal theft, especially in places previously untouched by the problem.[20] Benue Province was one of the areas where theft became prevalent for the first time. The deteriorating living conditions of blacksmiths, arising partly from the economic situation and raw-material shortage, accelerated the theft of railway sleepers and keys in Benue Province. As elsewhere, black-

15 A Railway Ordinance was introduced in the 1920s to counteract metal thefts but it failed to curb the problem.

16 There were large-scale thefts of railway sleepers and keys around the Dangora section of the railway in 1921. Eight persons were arrested and convicted: one man was sentenced to two years in prison, two men to one year, four men to six months and one man to nine months' imprisonment.

17 As Jaggar (1994: 35) pointed out, a shilling's worth of scrap metal could produce up to two dozen hoe blades, each selling for 5 to 6 pence.

18 The extensive sale of unserviceable scrap metal to blacksmiths started in the 1950s (Yusuf 2011b: 17) but new evidence suggests it may have started even earlier.

19 Freund (1982) pointed out that the thefts were a result of the ban on iron smelting.

20 The preceding years were a time of severe economic meltdown with imports and exports temporarily checked.

smiths were arrested and convicted in an attempt to curtail their activities (NAK SNP 17/2 119.59). Evidence shows that both railway and non-railway workers were involved in thefts.[21] As the Resident of Benue Province observed, '(i)ron stealing from the (r)ailway will (...) continue as long as the connivance and assistance of blacksmiths is obtainable' (*Ibid.*).

Imitation or innovation?

The raw-materials shortage and the structural changes emanating from the import restrictions during World War II resulted in technical and entrepreneurial innovations in blacksmithing. The ban on imports of European-manufactured consumer goods created an acute scarcity of manufactured metal products such as door bolts, hasps, hinges, shovels and head pans throughout the country and promoted demand for locally made metal products (Jaggar 1994). The supply of these products was equally restricted. The wartime scarcity of European manufactured goods had repercussions for a number of local crafts and there were shortages of iron and farm implements across the region.[22] For instance, in February 1945, the Resident of Benue Province reported that '(h)oes, knives etc. are in short supply' (NAK ZAR PROF 570) and in March 1945, a police officer remarked that 'steel is almost unobtainable these days and in consequence command a high price' (NAK ZAR PROF RLY/2). The termination of shipments from Europe created a scarcity of imported manufactured goods in the region[23] and stimulated demand for substitute products. This resulted in the local imitation of imported products by the local smithing industry.[24]

21 See also Ochonu (2009: 81-82) and Yusuf (2011a, 2011b).

22 As Koll (1982: 92) points out, smiths in Southern Nigeria became impoverished as a result of the post-war changes.

23 Kano (the region's commercial centre) and Kaduna (the regional headquarters) had large populations of Europeans (colonial officials, members of the trading community and allied soldiers) as well as educated African elites who expected European-style houses. Kano in particular (by virtue of its position as an air base for Allied contingents to the Middle East, India and Burma) required specially constructed European houses and barracks as well as European consumer goods.

24 A revival in smelting activities across Africa occurred between the 1920s and the 1940s. Jaggar (1994: 27) reports a revival of smelting in Zamfara in the 1940s following the ban on imports of European manufactured goods. Sasson observed a similar revival in smelting among the Sukur famers of northeastern Nigeria in the 1960s because of the low quality of imported iron (Pole 1982: 507). Herbert (1993: 11) also noted a revival of smelting in northern Ghana because of 'hard times'.

Local demand for imported manufactured goods increased demand for local imitations of imported metal products and blacksmiths turned import restrictions to their advantage by producing imitations of imported metal products to meet growing urban requirements. As with most changes, this was a difficult step for many smiths. Initially, only a few who realized the prospects stepped in to satisfy the growing demand (*Ibid.*). Blacksmiths showed a sense of entrepreneurship by charging higher prices for their products, knowing that they had no competition in the market. And although this was effective for some time, Jaggar noted that by the end of World War II, smiths had lowered their prices to compete with imports.[25]

The smiths managed to turn a disadvantageous position to their advantage by pioneering the local imitation of imported manufactured metal products. Smiths should be seen as innovators, given that they initiated the reproduction of foreign products even though the products were previously in use elsewhere (Jaggar 1994: 81). Katzin (1964: 183) described these smiths as innovating entrepreneurs, as they conceived an idea and introduced new elements into the economy. He used the term 'imitating entrepreneurs' to describe anyone who adopted a new economic practice after a local innovator had demonstrated it to be practical and profitable. Critics have argued that imitators cannot be considered as innovators because the idea emanates from elsewhere and not from within their own environment (Koll 1968). However, as Katzin points out, imitators too can be seen as innovators because, like innovators, they must be ready to integrate new elements into their environment or culture. Although it is not possible to provide information on the productive capacity of blacksmiths in this period, the evidence available suggests that railway metals, in the form of sleepers, keys, fish plates, spikes and even unserviceable scrap metal, were stolen in large quantities across the rural areas, particularly the southern part of Zaria and Kano Provinces and especially in the Dangora section (NAK ZAR PROF 570; NAK ZAR PROF RLY/2; NAK KAN PROF RLY/3; NAK ZAR PROF 4182 Zaria Province Annual Report for 1947).[26] If these thefts were in response to the demand for local imitations of manufactured imported metals, then demand was presumably even higher. Despite official attempts to curtail the practice, metal thefts persisted until the final decade of British colonial rule. Large numbers of

25 Locally made 10-inch bolts that sold for 60 shillings during the war were cut to 24 shillings per dozen at the end of the war.

26 Unserviceable scrap metals were accumulated as a result of the process of upgrading the railway system over the years.

railway keys were stolen in Kano Province and along the Kano-Nguru line (NAK RLY/3) and, in the long run, the British distributed the scrap railway sleepers and keys to blacksmiths across the region (see Table 14.1). Previously, it was assumed that scrap metal was first distributed to blacksmiths in the 1950s (Yusuf 2011a, 2011b)[27] but distribution may in fact have started in the 1940s or even earlier. For example, some older sleepers were found during a search of blacksmiths' properties in Zaria Province in 1945. When asked where they were procured, the blacksmith produced a written permit issued by the railway authority relating to the purchase of the said sleepers and other types of 'railway irons'. This suggests that the distribution of railway scrap metal to the Native Authority must have started by that time or earlier (NAK ZAR PROF RLY/2).[28]

Table 14.1 Railway scrap metal distributed to blacksmiths across Northern Nigeria, 1954-1955

Province	Native Administration	Quantity (no. of items)
Adamawa	Muri	3,048
	Adamawa	1,000
Bauchi	Katagum	300
	Misau	100
	Bauchi	3,000
Borno	Borno	1,000
	Fika	Value of £100 including freight
	Bedde	Value of £50 including freight
	Biu	Value of £40 including freight
Ilorin	Ilorin	1,000
	Pategi	80
	Lafiagi	120
	Kaima	200

27 The scrap metals were distributed directly to the Native Authority across the region, subject to the blacksmiths' demands. The railway played an important role in distributing the metals across the region, preferring to offload the scrap at the urban and peri-urban railway towns. At times, they were distributed via the railway motor service. Offloading scrap outside the approved distribution points was done by the blacksmiths themselves.

28 Station Master Zonkwa stated that the sleepers were bought in the Lagos and Kano areas by a one Mai Tsabega who dispatched them to a M. Musa Na Yar-mallam at Zonkwa village to sell. Nearly all blacksmiths from this area obtained their iron from this man.

Table 14.1 (*cont.*)

Province	Native Administration	Quantity (no. of items)
	Bussa	250
Kabba	Kabba	200
	Yagba	500
	Igala	500
	Bassa Komo	50
	Kwarra	60
	Igbira	2,000
Kano	Kano	10,000
	Gumel	350
	Hadejia	180
	Kazaure	200
Katsina	Katsina	1,200
	Daura	400
Niger	Bida	400
	Agaie	200
	Kontagora	80
	Zuru	200
	Gwari	100
	Kamuku	40
	Abuja	200
	Lapai	150
Plateau	Jamaa	200
	Pankshin	200
	Kanam	150
	Lowland	150
	Jos	700
	Southern	450
Sokoto	Sokoto	900
	Sokoto	1,100
	Gwandu	1,700
	Argungu	160
Zaria	Zaria	1,100
	Birnin Gwari	100
	Kagoro	200
	Jaba	200
	Moroa	100

Source: National Archives Kaduna (NAK) RLY/15 Distribution of scrap sleepers by Nigerian Railway Corporation, 1954-1957

What characterized smithing in this period was the remarkable innovation and entrepreneurship in the sector. This resulted in the emergence of a new craft, and this industry came to be known as 'cold' or 'white-smithing', as opposed to hot, fire or blacksmithing, which is the original method of smithing.[29] The technology of coldsmithing was adapted to imitate manufactured European metal products. It was essentially a light-weight craft technology as few of the tools and technical processes of real smithing are used in coldsmithing. Initially, when coldsmithing was developed, imitations of imported metal products like door bolts, hasps, hinges, shovels and head pans were manufactured. The existing literature on blacksmithing attests to the 'newness' of coldsmithing in Hausaland (Wolff 1986; Aliyu *et al.* 2008). A similar distinction is made among Yoruba blacksmiths in Southern Nigeria and the Adar smiths in Niger where work in black-metal iron is distinguished from work in white-metal silver, aluminium and brass (Lloyd 1953; Pilaszewicz 1991).

This new craft has attracted many new entrants and patrons over the years due to its reasonable price, the availability of scrap metal, its malleability and versatility. Whereas the technical process in blacksmithing requires more than one man, coldsmithing is relatively simple and requires only one-man production units. Coldsmithing can be set up anywhere provided that the necessary tools, such as a light-weight hammer, chisel, flat anvil and some capital, are available (Jaggar 1994: 69). Although it is not clear where the idea of coldsmithing came from and how it spread to Northern Nigeria, it has become a popular craft. The range of products manufactured by coldsmiths has increased from basic door bolts, hasps, hinges, shovels and head pans to include other items like mouse traps, trowels, metal cooking trivets, watering cans, piggy banks, kerosene tins, oil drums and bicycle stands. Coldsmiths were always on the lookout for new and popular consumer products (Heathcote 1976; Jaggar 1994). The range of products manufactured by coldsmiths was determined by the type of scrap metals available and nowadays metal workers make use of scrap metals such as car parts, mild steel, and cast iron from piston rings and aluminium as well as iron rods. Coldsmiths were always searching for imported products that could be adapted and reproduced locally, which is evidence of their innovativeness and entrepreneurship (Jaggar 1994; Heathcote 1973).

This new substitutive industry enjoyed a relative boom as smiths charged high prices for their products. The import restrictions following World War II conferred a monopoly on them over their products and they exploited this to their own advantage. By the end of the war, a number

[29] A major difference between the two is that while coldsmithing involves hammering to produce implements at a low temperature, blacksmithing involves hammering at high temperatures.

of pioneering coldsmiths had accumulated a lot of capital with which they charted the course of the industry over the following decades. The coldsmiths' craft experienced further growth during the Nigerian Civil War when imported manufactured metal products were in short supply because of the ban on imports and, when available, prices were unbearable. The price of imported door locks rose from 15 shillings in 1967 to 30 shillings in 1968. The import restriction was a boost to the craft of coldsmithing. It stimulated the imitation of imported manufactured metal products as it had done during World War II (Jaggar 1994: 64-87).

If the interwar period was a time of experimentation and innovation in coldsmithing, the end of the Nigerian Civil War was a period of marked transformation, from production by hand to modern, industrial production by machine. Profits accumulated over the years meant that there was a lot of cash in the hands of the coldsmiths who came to pioneer the establishment of a modern industrial metal-working industry. This development started in Kano among the city's blacksmiths and the idea of establishing an industry was conceived in 1969 by the same group of smiths who pioneered coldsmithing. Due to the enormous capital requirements involved in establishing a modern industry, which the smiths were unable to raise, they sought assistance from Kano's state government. The state, which had been established in 1967, was attracted by the proposal, especially because it fell within the government plan of setting it on an industrial path. The government approved the sum of £4,500 out of the total of £15,000 of floating capital that was required for the industry's take-off. This loan was to be paid back within a period of eight years, at an annual interest rate of 5%, with the possibility of a further loan. The proprietors raised £3,000 among themselves, with the remainder coming from local entrepreneurs who acquired shares in the company. While this innovation was applauded as a landmark in the traditional craft industry, people also expressed their concerns and fears about the industry's survival, especially in the face of competition from imported manufactured goods. The company eventually closed down in the 1980s, although it was later revived and renamed Madakin Kira Trading Company. It is currently involved in producing aluminium, steel and metal products (Jaggar 1994: 88-99).

Conclusion

The blacksmithing industry has undergone tremendous transformations that shaped both the type of raw materials blacksmiths used and the technology of the craft itself. British colonialism, though not the only factor, was indeed a transformative force in these developments. British colonial policies accelerated the decline of rural mining and smelting, thereby starving rural smiths of the raw materials they required. Blacksmiths resorted to stealing from the railways to compensate for the shortage of raw materials they were facing. Restrictions on imported manufactured metals were a major boost to the expansion of smithing, leading to technical innovations that led to the emergence of coldsmithing within the smithing sector. Further developments within coldsmithing also resulted in the transformation of smithing from being a traditional craft driven by hand to a modern industry powered by machines. This chapter has attempted to show how smiths were the agents of change and innovation, even though the changes were influenced by circumstances triggered by British colonialism.

References

Aliyu, A.B., H. Yunusa & A.A. Adamu (2008) 'Investigative Study of Archaeo-Metallurgy of Katsina State', *Continental Journal of Engineering Sciences* 3(12): 1-12.

Anderson, D. (1986) 'Stock Theft and Moral Economy in Colonial Kenya', *Africa, Journal of the International African Institute* 56(4): 399-415.

Callaway, A. (1967) 'From Traditional Crafts to Modern Industries'. In: P.C. Lloyd, A.E. Mabogunje & B. Awe, eds, *The City of Ibadan: A Symposium on its Structure and Development*, Cambridge: Cambridge University Press, pp. 153-171.

Cline-Cole, R.A. (1994) 'Political Economy, Fuel Wood Relations and Vegetation Conservation: Kasar Kano, Northern Nigeria, 1850-1915', *Forest and Conservation History* 38(2): 67-78.

Darling, P.J. (1989) 'Archeology and the Dating of Historical Events in Kano'. In: B.M. Barkindo, eds, *Kano and Some of Her Neighbours*, Zaria: Ahmadu Bello University.

Emeagwali, G.T. (1997) 'Colonialism and Science in Africa by 1914'. In: H. Selin, ed., *Encyclopedia of the History of Science and Medicine in Non-Western Cultures*, Alphen a/d Rijn: Kluwer Academic Publishers.

Freund, B. (1982) 'Thefts and Social Protest among the Tin Miners of Northern Nigeria', Paper presented at the African Studies Programme at University of Illinois, 21-24 April.

Goucher, C.L. (1981) 'Iron Is Iron 'til It Is Rust: Trade and Ecology in the Decline of West Africa Iron Smelting' *The Journal of African History* 22(2): 179-181.

Heathcote, D. (1976) *The Arts of the Hausa*, London: World of Islam Festival Publishing Company.

Herbert, E.W. (1993) *Iron, Gender, and Power: Ritual of Transformation in African Societies*, Bloomington: Indiana University Press.

Hill, P. (1972) *Rural Hausa: A Village and a Setting*, Cambridge: Cambridge University Press.

Hogendorn, J.S. (1978) *Nigerian Groundnut Exports: Origin and Early Development*, Zaria: Ahmadu University Press.

Hopkins, A.G. (1973) *An Economic History of West Africa*, London: Longman.

Jaggar, J.P. (1973) 'Kano City Blacksmiths: Pre-colonial Distribution, Structure and Organization', *Savanna* 2(1): 11-26.

——. (1994) *The Blacksmiths of Kano City*, Cologne: Rudiger Koppe Verlag.

Katzin, M. (1964) 'The Role of the Small Entrepreneur'. In: M.J. Herskovits & M. Harwitz, eds, *Economic Transition in Africa*, Evanston: Northwestern Uiversity Press, pp. 179-198.

Koll, M. (1969) 'Craft and Cooperation in Western Nigeria', Freiburg: Bergstraesser Institute.
Lloyd, P. (1953) 'Craft Organization in Yoruba Towns', *Journal of the International African Institute* 23(1): 30-44.
Lugard, L. (1965) *The Dual Mandate in British Tropical Africa*, London: Frankass.
Mabogunje, A.L. (1968) *Urbanization in Nigeria*, London: University of London Press.
Nalado, S. (2004) 'A History of Anka Metal Industry: 1815-1899', Unpublished MA Thesis, Bayero University, Kano.
Ogundiran, A. (2005) 'Manufacturing Indigenous'. In: Zelza, P.T. & D. Eyoha, eds, *Encyclopedia of Twentieth-Century African History*, London: Routledge, pp. 347-350.
Okafor, E.E. (1993) 'New Evidence on Early Iron Smelting from Southeastern Nigeria', in: Thurstan, S., P. Sinclair, B. Andah & A. Okpoko, eds, *The Archeology of Africa: Food, Metals and Towns*, London: Routledge, pp. 432-448.
Okediji, F.A. (1972) 'An Economic History of Hausa Fulani Emirates of Northern Nigeria: 1900: 1939', Unpublished PhD Thesis, Indiana University.
Pfaffenberger, B. (1992) 'Social Anthropology of Technology', *Annual Review of Anthropology* 21: 491-596.
Pilaszewicz, S. (1991) 'The Image of Hausa Smiths in Some Written Sources'. In: Monino, Y., ed., *Forge et forgerous*, Paris: Orstrom.
Pole, L.M. (1982) 'Decline or Survival? Iron Production in West Africa from the Seventeenth to the Twentieth Centuries', *Journal of African History* 23(4): 503-513.
Shea, P.J. (1974/1977) 'Economics of Scale and the Indigo Dyeing Industry of Pre-colonial Kano', *Kano Studies* 1(2): 55-61.
Wolff, N.H. (1986) 'A Hausa Aluminum Spoon Industry', *African Arts* 19(3): 40-44.
Yusuf, S.T. (2011a) 'The Incidences of Theft of Railway Metals in Colonial Northern Nigeria', Paper presented at the 9th International Conference of the International Association for the History of Transport Traffic and Mobility held at the Deutsches Techniksmesum, Berlin, 6-9 Otober 2011.
——. (2011b) 'The Incidences of Theft of Railway Metals in Colonial Northern Nigeria', *Afro Asian Journal of Social Sciences* 2(4): 1-23.
——. (2011c) 'The Great Longing for Railways: How the Baro to Kano Railway Connected Northern Nigeria with Centres of Industrialization', Paper presented at the 5th International Railway History Association (IRHA) Conference at the Centre for Urban History of East Central Europe Lviv, Ukraine, 3-5 November.

Archival materials

KSHCB SNP/9/2/1924, Kano Province Annual Report, files 635
KSHCB MLG 16490/1932, Kumbotso District Kano Emirate Reassessment
NAK MLG 9117/1929, Kura District of Kano Emirate Reassessment
NAK ZARPROF 2561/1914, Zaria Provincial Annual Report 1914
NAK NAK ZARPROF 2562/1915, Zaria Provincial Annual Report 1915
NAK NAK ZARPROF 2475/1917, Zaria Provincial Annual Report 1917
NAK ZARPROF 163/1919, Zaria Provincial Annual Report 1918
NAK 2572/1925, Zaria Provincial Annual Report for the Year 1925
NAK ZARPROF 4182, Zaria Annual Report 1947s
NAK ZAR PROF RLY/2, Offences on Trains and Railway Stations
NAK KAN PROF RLY/3, Theft of Railway Materials
NAK ZARPROF 570, Theft of Railway Materials 1947-56
NAK 299/S.3, Vol. 1 Railway Publications 1949-56
NAK RLY/15, Distribution of Scrap Sleepers by Nigerian Railway Corporation, 1954-1957

List of authors

Stefano Bellucci is a researcher at the International Institute of Social History in Amsterdam where he also coordinates the preservation and collections activities of labour and social history in Sub-Saharan Africa. He previously taught African history and politics at the University of Pavia in Italy. He has a PhD from the Faculté Jean Monnet de Droit Économie et Géstion at the University of Paris XI and an MA in political science from Miami University. He is currently researching the history of labour in Eastern Africa and the Horn of Africa.
sbe@iisg.nl

Marleen Dekker is a development economist and human geographer at the African Studies Centre in Leiden where she is researching the role of social networks in accessing (informal) insurance, markets and local socio-economic development. Risk and coping strategies are central in her work on land reform and the economic crisis in Zimbabwe, intra-household bargaining in Ethiopia and community-based health insurance in Togo. She recently co-edited *Markets of Well-being: Navigating Health and Healing in Africa* with Rijk van Dijk (Brill, 2010).
Mdekker@ascleiden.nl

Gregor Dobler is Professor of Social and Cultural Anthropology at Freiburg University in Germany where his research focuses on economic and political issues in Southern Africa and France. He is writing a book on traders in colonial Ovamboland entitled *Traders and Trade in Colonial Ovamboland, 1925-1990. Elite Formation and the Politics of Consumption under Indirect Rule and Apartheid* (Basel Africa Bibliographies) that will be published in 2013 and on which his chapter in this current volume is based.
gregor.dobler@ethno.uni-freiburg.de

Dick Foeken is a senior researcher specializing in urban poverty, urban water supply and urban agriculture at the African Studies Centre in Leiden. He has done extensive field research in Kenya and other countries

on food security, land policy and urban agriculture. He has published widely on issues related to urban agriculture and political geography in *Political Geography, The Journal of Peasant Studies, Food and Nutrition Bulletin, Ecology of Food and Nutrition, Habitat International, Geoforum* and *GeoJournal* and has co-edited three volumes in the ASC's *African Dynamics* series.
dfoeken@ascleiden.nl

Jan-Bart Gewald is a historian and senior researcher at the African Studies Centre in Leiden. He has published on a wide range of subjects in African history and is specifically interested in the history of the social relationship between people and technology in Africa. His publications include *Herero Heroes: A Socio-political History of the Herero of Namibia 1890-1923* (James Currey, 1999) and *The Speed of Change: Motor Vehicles and People in Africa, 1890-2000* (Brill, 2009), which he edited with Sabine Luning and Klaas van Walraven.
gewald@ascleiden.nl

Bill Kinsey is a senior research fellow at the Ruzivo Trust in Harare and also has an honorary appointment at the Centre for the Study of African Economies at the University of Oxford. With graduate degrees in economics, anthropology and nutrition from Stanford University, his research interests focus on empirical policy analysis of issues related to rural poverty, including the long-term welfare dynamics of rural households, land-reform programmes in Southern Africa and elsewhere, the welfare effects of state policies on rural households, child health and nutrition, and resource management. These are being addressed in the Zimbabwe Rural Households Dynamics Study, an ongoing study of 550 farm households that he initiated in 1982.
bkinsey@mango.zw

Peter Knorringa is Professor of Private Sector and Development at the ISS, Erasmus University in Rotterdam. His present research focuses on the role of private-sector actors in development processes, particularly on the developmental relevance of CSR, standards and alternative economic development initiatives. He has published extensively on small and medium-scale enterprises, industrial clusters, global value chains and the role of trust in industrial development in *Development & Change, Enterprise Development & Microfinance, European Journal of Development Research, Journal of Development Studies* and *World Development.*
knorringa@iss.nl

André Leliveld is a development economist at the African Studies Centre in Leiden where he does research on informal insurance and social-support systems and community-based (health) insurance schemes in Africa, in particular in Togo and Uganda. His recent research interests also include structural transformation processes and innovation in African economies. He co-edited *Inside Poverty and Development in Africa: Critical Reflections on Pro-Poor Policies* with Marcel Rutten and Dick Foeken (Brill, 2008).
Leliveld@ascleiden.nl

Moses Mwangi is a lecturer at South Eastern University College's School of Water Resources, Science and Technology at the University of Nairobi where he also chairs the Department of Hydrology and Water Resources Management. His research interests include integrated water resources management and development. He has been involved in many projects over the last 25 years, particularly those supporting people's social, political and economic needs in arid and semi-arid lands. He has worked in various positions in research, resource development and the management of livelihoods in Kenya and internationally.
Mnmwangi2000@yahoo.co.uk

Paul Nugent is Professor of Comparative African History and Director of the Centre of African Studies at the University of Edinburgh. He has worked on the history and political economy of borderlands in West Africa and the electoral politics of Ghana. While holding a fellowship at the Stellenbosch Institute of Advanced Study (STIAS), he has been writing a book on the history of the South African wine industry in the twentieth century based on research funded by the Leverhulme Trust. He is currently President of AEGIS, the European Association of African Studies.
Paul.Nugent@ed.ac.uk

Samuel O. Owuor is a senior lecturer and chairman of the Department of Geography and Environmental Studies at the University of Nairobi, Kenya, where he has been teaching urban geography since 1995. He has a PhD in urban geography from the University of Amsterdam and has published in *Geoforum, Development Southern Africa, Flux* and *Hekima*. He has written a number of book chapters and is the author of *Bridging the Urban-Rural Divide: Multi-spatial Livelihoods in Nakuru Town, Kenya* (ASC, 2006).
samowuor@uonbi.ac.ke

Iva Peša is a historian and is currently doing her PhD at Leiden University on the social history of Mwinilunga District in northwestern Zambia. Her interests include the changing patterns of (agricultural) production, consumption, labour migration and social relationships.
i.pesa@hum.leidenuniv.nl

Marcel Rutten is a geographer and researcher at the African Studies Centre, Leiden, and Radboud University, Nijmegen. Since the early 1980s he has done longitudinal research on land tenure, water development and food security in semi-arid areas of Africa and his current research is on sustainable water projects, natural-resource conflicts and the individualization of land ownership in Kenya. His publications include *Out for the Count: The 1997 General Elections and Prospects for Democracy in Kenya* (Fountain, 2001, with A. Mazrui & F. Grignon) and *Inside Poverty and Development in Africa: Critical Reflections on Pro-poor Policies* (Brill, 2008, with André Leliveld & Dick Foeken).
rutten@ascleiden.nl

Cees van Beers is an economist and Professor of Innovation Economics and Management in the Department of Technology, Policy and Management at Delft University of Technology. His research focuses on two main themes: the determinants of innovations in firms and the impact on their performance; and the impact of government policies and regulations on fossil-fuel energy use. He has published extensively on these topics in *Research Policy, Journal of Innovation Product Management, Industrial and Corporate Change, Technology Analyses and Strategic Management, Small Business Economics, Ambio* and *Applied Economics.*
C.P.vanBeers@tudelft.nl

Rijk van Dijk is an anthropologist at the African Studies Centre, Leiden where he has done extensive research on the rise of Pentecostal movements in the urban areas of Malawi, Ghana and Botswana. His current research deals with religious, particularly Pentecostal, engagements with the domains of marriage, sexuality and HIV/AIDS in Botswana. His publications include 'Gloves in Times of AIDS: Pentecostalism, Hair and Social Distancing in Botswana', in F. Becker & P.W. Geissler (eds), *AIDS and Religious Practice in Africa* (Brill, 2009) and he is the editor-in-chief of *African Diaspora: A Journal of Trans-national Africa in a Global World* that is also published by Brill.
dijkr@ascleiden.nl

Klaas van Walraven is a political scientist and historian at the African Studies Centre in Leiden where he does research on West Africa, in particular Niger. His extensive study on the history of Niger in the 1960s and 1970s entitled *The Yearning for Relief: A History of the Sawaba Rebellion in Niger, 1954-1974* will be published by Brill in 2013 and will also be translated into French. He is one of the editors of the *Africa Yearbook: Politics, Economy and Society in Sub-Saharan Africa* (Brill).
walraven@ascleiden.nl

Shehu Tijjani Yusuf is an assistant lecturer at Bayero University, Kano where he is researching the history of the impact of railroads on Northern Nigeria from 1908 to the 1970s. His recent publications include 'Towards a New Approach to Nigeria's Railway History: The Rural and Agricultural Alternatives', in S. Aderinto & P. Osifodunrin (eds), *The Third Wave of Historical Scholarship on Nigeria: Essays in Honor of Ayodeji Olukoju* (Cambridge Scholars, 2012); 'The Incidences of Theft of Railway Metals in Colonial Northern Nigeria', *Afro Asian Journal of Social Sciences* 2:24 (2011); and *The Impact of the Railway on Kano Emirate, c. 1903-1960s: The Case of Madobi and Kwankwaso Towns* (Lambert, 2010).
stijjani@gmail.com

Massimo Zaccaria is a senior researcher in African and Islamic Studies in the Department of Social and Political Sciences at the University of Pavia in Italy where he specializes in the history of the Horn of Africa during the colonial period. He has a PhD in African History from the University of Siena and is the author of *Le note del commissario. Teobaldo Folchi e i cenni storico amministrativi sul commissariato di Massaua* (Franco Angeli, 2009).
zaccaria@unipv.it

* 9 7 8 9 0 0 4 2 4 5 2 3 5 *